Introduction to Physically Realizable Physics

Francis T.S. Yu

Copyright © 2024 Francis T.S. Yu

All rights reserved.

ISBN: 9798877098312

Understanding our temporal universe means having an inkling of the "distance" between where you are today and where you were yesterday.

CONTENTS

Preface vii
CHAPTER 1: Discovery of Temporal (t > 0) Universe 1
CHAPTER 2: The Fate of Schrödinger's Cat 15
CHAPTER 3: What is "Wrong" with Theoretical Physicists 19
CHAPTER 4: From Schrödinger's Equation to Quantum Conspiracy 29
CHAPTER 5: The Limits of Einstein's Theory of Relativity 39
CHAPTER 6: Quantum Qubit Information Conspiracy 49
CHAPTER 7: Why modern physics is so weird… and so wrong 57
CHAPTER 8: Does Einstein's General Theory Belong to the Realm of Science? 65
CHAPTER 9: Einstein's Spooky Distance 78
CHAPTER 10: Can Space Really Curves Spacetime? 85
CHAPTER 11: Dark-Age of Our Modern Physics 97
CHAPTER 12: Where Dark Matter and Dark Energy comes from? 106
CHAPTER 13: Why Einstein's relativistic mechanics is against the laws of nature. 121
CHAPTER 14: Myth of Entropy - Boltzmann's Exorcist 131
CHAPTER 15: Schematically Disproves Bogus Modern Principles 148
CHAPTER 16: From classical to physical realizable Hamiltonian Mechanics 162
CHAPTER 17: Enigma of Eigen state wave function 179
CHAPTER 18: Why mathematics is not science 192
CHAPTER 19: Origin of our science 205
CHAPTER 20: Origin of de Broglie wave dynamics 213
CHAPTER 21: Interpretation of $\psi(t)$ 222
CHAPTER 22: Why ancient Chinese did not have a fundamental science 234
CHAPTER 23: AI is at different level of knowledge abstraction 239
CHAPTER 24: Theory of temporal (t > 0) Space 244
CHAPTER 25: Dirac's antimatter hypothesis is not science 255
YouTube Links for narration by Chapters 263
About the author 267

Preface

Science is rooted in our ability to observe the world, either directly with our own senses, or indirectly with the help of instruments. It follows that scientific laws, theories and principles arise from our collection of data about the world (or to put it more plainly, our observation of it), and the conclusions and generalizations we draw from this data. There are of course, exceptions to this rule, and it is in the exceptions that we tend to encounter problems. In modern physics, for example, where observation can take a back seat to manipulating mathematical symbols on either side of the "=" sign, many leaders in the field have become more reliant on equations than on observable reality. As a result, some of the constituent laws, theories, and principles in theoretical physics are rooted in assumptions that are either difficult to impossible to prove/disprove or have simply been taken for granted by the scientific community at large, and consequently overlooked for years to decades. Obviously, this presents a problem for any scientist looking to be, well... scientific.

It is important to remember that for a law, theory, or principle to be tenable, it should fulfill both of the following two prerequisites: 1) accord with the rules of mathematics and/or formal logic; 2) be rooted in assumptions that accord with (provable) material reality. Why? Because theories developed in the absence of logic are, by definition, devoid of reason, while those rooted in logic, but not in material reality are more hypothesis than theory. Put more simply, if a theory is wholly illogical, then it is essentially a collection of non-sequitur statements; if it is solely dependent on logic and at the same time completely independent of physical evidence, or even the possibility of attaining the latter, then it is more faith-based than reason-based.

What advocates of modern physics seem to have forgotten is the simple fact that math is not science, even if science can and often does make use of math to further its own development. As I stated above, science could be considered the process of making increasingly accurate and precise generalizations about the physical world based on either direct or indirect observations of it. Mathematics, by contrast, could be considered the process of using logic to discover new ways of using, grouping and/or combining symbols. And even though some of these symbols may in some cases be used to represent some part of physical reality, they are not, in and of themselves physical reality. Thus, while science ideally maintains some tangible and direct connection to material reality via the subject's observation of it, mathematics normally lies at least one step removed from material reality.

Obviously, this does not mean that we cannot use mathematics to help solve scientific problems, or indeed, problems that are bound to arise in any given field. What it does mean, however, is that, as with any other tool, mathematics has limits. Some of these limits are revealed in the fact that the more interpretative work we must undertake to understand or explain how any group of symbols reflects some portion of material reality, the greater our chances of misinterpreting the information presented to us and in the process, bending material reality to fit our theory. This is why the more we rely on mathematics to generate new theories yet do so without a commensurate backing of observable evidence, the further we may end up straying from material reality and strolling therefore, into domains traditionally reserved for science fiction and religion. For just because mathematics is rooted in logic and reason does not mean that everyone's application or interpretation of it will be entirely reasonable.

History is riddled with examples of theories that were once widely accepted to the point of becoming "common knowledge," yet are today considered disproven forms of pseudoscience. All such examples could be a result of theorists of any given era lacking in sufficient real-world feedback—material evidence that if widely available and accepted at the time such theories were emerging, could have prevented science from straying so far from reality. In all such cases, one could say that said theorists went as far as to bend reality to fit their theory rather than vice versa—no minor problem, if we consider the former to be how science fiction works, and the latter to be the way science (non-fiction) should work. Likewise, today's scientists, particularly in the fields of modern physics and quantum computing, bend time, space, and other aspects of material reality in order to fit them into their equations. It follows that overreliance on mathematics in these fields has led, not only to a spate Hollywood flicks that have, over the decades, depicted time travel, along with a slew of other fantastical myths, but more importantly, to a spate of (quasi-)scientific theories presented in the form of books, articles, and talks, that back up more than a few of these myths. Hollywood producers, screenwriters and directors, whose job it is to transport audiences into the realm of make-believe, can be forgiven for conflating science with fiction. Members of the scientific community, however, have a different job description—namely, one of upholding observable reality (as opposed to denying it)—and should consequently be held accountable for what amounts to a major derelict of duty.

This book is a compilation of PowerPoint presentations I recorded over the last few years with the help of several friends and family members who so generously served as narrators. These presentations are my attempt to elucidate some of the fundamental problems that continue to plague modern physics, quantum computing and other fields that have become more rooted in the symbology of complex mathematics than in material reality. Each presentation is therefore meant to challenge some of the fundamental assumptions of modern physics and quantum computing and thereby help the viewer/reader differentiate between science and fiction.

Francis T.S. Yu
San Diego, CA, USA
January 2024

CHAPTER 1

Discovery of Temporal (t > 0) Universe

Dream of a spacetime continuum…

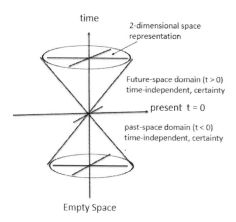

…created a virtual spacetime continuum!

The following is an article published in the Asian Journal of Physics. It also appears in one of the chapters of Francis Yu's recently published book, *Entropy, and Information Optics: Connecting Information and Time.*

This article is based on the constraints of the laws of physics to illustrate time (variable) as the origin of creating our physical space (i.e., our universe).

Physical reality exists with the passing of time, and it is supported by laws of science.

Virtual reality is created without the constraint of time, and it is not supported by laws of physics.

F. T. Yu, *"Time: The Enigma of Space," Asian Journal of Physics*, Vol. 26, No. 3, 149-158, 2017.

Time

One of the enigmatic variables in the rules of science is "Time". So, what is time?

Time is a variable, and it has no substance, it has no mass, no weight, no coordinate, no origin, and it cannot be detected or even be seen. It only can move forward, and it cannot move backward.

Yet, time is everlasting. Everything is associated with time; this includes our universe.

When dealing with science, time is one of the most intriguing variables that we cannot simply ignore.

Strictly speaking, all the laws of physics as well all physical substances could not exist without the existence of time.

Energy & Energy Reservoir

Energy is a physical quantity that governs the existence of substances.

Every substance is created by energy and every substance can convert back to energy. Therefore, Energy and Substance are exchangeable.

Mass can be treated as an "energy reservoir".
Our universe is compactly filled with mass and energy.
Nonetheless, without the existence of time, the trading between mass and energy cannot happen.

Einstein's Energy Equation

Einstein's Energy Equation as given approximately by.

$$\varepsilon \approx mc^2$$

where Energy ε is approximately equal to rest mass m, c is the velocity of light.

Since all of the laws in science are approximations, I intentionally used the approximation sign.

However, in practice, the amount of energy produced should be smaller than the equivalent amount of mass as shown by the following equation:

$$\varepsilon < mc^2$$

Energy Conversion

In view of the Einstein's equation, it takes time to convert mass into energy. Thus, without the inclusion of time variable, the conversion would not get started. In fact, Einstein's equation is a representation of total amount of energy to be converted from a rest mass.

In other words, every mass has an energy storage capacity, as given by the Einstein's equation.

By incorporating the time variable, Einstein equation can be represented by a partial deferential form as given in the equation below:

$$\frac{\partial \varepsilon(t)}{\partial t} = c^2 \frac{\partial m(t)}{\partial t}$$

where the partial derivative of energy with respect to time

$$\partial \varepsilon(t)/\partial t$$

is the rate of energy conversion, c is the speed of light and the partial derivative of mass with respect to time?

$$\partial m(t)/\partial t$$

is the corresponding rate of mass reduction.

Trading Mass & Energy

One of the important aspects of Einstein's energy equation is that energy and mass can be traded. So that the rate of energy creation from mass can be written in terms of Radiant Energy as given by the equation below:

$$\frac{\partial \varepsilon}{\partial t} = c^2 \frac{\partial m}{\partial t} = (\nabla \cdot S) = -\frac{\partial}{\partial t}(\frac{1}{2} \epsilon E^2 + \frac{1}{2}\mu H^2)$$

In which we see that the exploded Energy (i.e., radiation) diverges from mass (i.e., $\nabla \cdot S$) del dot S is the energy vector, as mass decreases with time. And the last term represents the diverging radiation energy density.

Similarly, the creation of mass from energy can also be shown by the following equation:

$$\frac{\partial m}{\partial t} = \frac{1}{c^2}\frac{\partial \varepsilon}{\partial t} = -\frac{1}{c^2}(\nabla \cdot S) = \frac{1}{c^2}\frac{\partial}{\partial t}(\frac{1}{2} \epsilon E^2 + \frac{1}{2}\mu H^2)$$

The major differences of this equation, as compared with former one, is the energy convergent operator (i.e., $-\nabla \cdot S$) minus del dot S. In which we see that energy converges into a small volume to create the mass.
Since mass creation is inversely proportional to the speed of light squared c^2, it will take a lot of energy to produce a small quantity of mass.

(m). Nevertheless, in the cosmological environment, availability of huge amounts of energy has never been a problem.

Physical Substance

In our physical world, every matter is a substance. The reason is that they were all created by means of energy or mass. As we mentioned before, our physical space (i.e., our universe) is fully packed with substances (i.e., mass and energy); there is no absolute empty space in it.

Nonetheless, all physical substances must exist with time; in our universe no physical substance can exist forever or exists without time.

Thus, every physical substance is involved with Time and Space, which I call Temporal Space.

Therefore, we see that, without time there would be no substance and no universe.

Physical substance and Time

On the other hand, time cannot exist without the existence of substance. That is time and substance must simultaneously exist; one cannot exist without the other.

Therefore, if our universe must exist with time, then our universe will eventually get old and die.

So, the essence of time may not be as simple as we think: For example, as species live in a far distant galaxy move closer to the speed of light, their time goes somewhat slower relatively to ours.

We can see that time may not be the same at different subspaces, for example at the edge of our universe.

As we know every conversion (either mass to energy or energy to mass) cannot start without the ignition of time. Then time and substance (i.e., energy and mass) must simultaneously co-exist. So, every physical substance, including our universe co-exist with time; we can describe the relationship by a Time-Space Dimensional representation as given by the following equation:

$$f(x,y,z;t), t > 0$$

This equation is a function of space and time, where time is a forward variable.

Substance & Physical space

Absolute empty space cannot exist in physical reality.

Every physical space needs to be filled with substances, (includes our universe), it is not possible that there is any empty space in it.

Let me repeat mass (and energy) co-exists with time. As we said before without the existence of time there would be no mass, no energy and no universe.

We see once again that every physical substance co-exists with time.

Our universe is a Bounded Subspace

Since a physical space cannot be imbedded within an absolute empty space, then our universe must be imbedded in a more complex physical space. This of course remains to be found.

If we accepted our universe is imbedded in a more complex space, then our space must be a bounded subspace.

It is our aim (within the laws of science) to investigate the enigma of time as it exists within our universe.

Back to Einstein's energy equation as given in differential form,

$$\frac{\partial \mathcal{E}}{\partial t} = c^2 \left(\frac{\partial m}{\partial t}\right) = \pm (\nabla \cdot S)$$

Where positive and negative del dot S [i.e., $\pm (\nabla \cdot S)$] represents the Divergent and the Convergent Energy operations, respectively. For example, black holes are the known celestial energy convergent operators, while stars (e.g., our sun) are the known divergent energy sources.

Moreover, the essence of the above equation shows the transformation, from dimensionless (i.e., singularity) quantity to dimensional (i.e., space-time) representation.

From this equation we see that, all the physical substances were created by energy and time, I.e., every physical subspace is created by substances. In short, we see that, del dot S is a function of space and time, and t is a time variable.

$$\nabla \cdot S = f(x,y,z;t), \quad t > 0$$

This is a time-space or a temporal-space representation.
Let us sidetrack a little and take a creative way to look at an equation.
An equation is not just a mathematical formula, it is…
  a symbolic representation,
  a description,
  a language,
  a picture,
  or even a video…

In view of the equation, we have developed:

$$\nabla \cdot S = f(x,y,z;t), \quad t > 0$$

del dot S equals to a function of Space and Time representation (where S is an energy vector and Time is a forward variable); we see that our temporal universe was created from a dimensionless singular explosion to a time-space dimensional representation!

In fact, our universe is still expanding currently, as based on current Hubble space telescope observation.

Time & Physical Space

One of the most intriguing questions in our life is the existence of time. Let me pose a question, if time does not exist then how do we know here is a physical space?

Thus, in the following we will see that there is a profound connection between time and space: If there is no time, then there will be no space!

A physical space in fact is a Temporal Space, in contrast to a Virtual Space.

Temporal Space is a space that is described by Time while Virtual Space is an imaginary space without the constraint of time.

In other words, every temporal space is supported by the laws of science, while virtual space may not be.

Trading time for space

A TV displayed image is a typical example of trading time for space, e.g., any TV displayed image of (dx, dy) (i.e., distant of x by distance of y) takes a section of time Δt to display it.

Since time is a forward moving variable, it cannot be traded back at the expense of space (dx, dy).

Therefore, it is the time that determines the physical space, and it is not the physical space that can bring back.

The amount of time Δt that had been used.

Ant it is the size (or dimension) of space that determines the amount of time Δt needed to create the space.

Speed of Light

Based on current laws of science, speed of light is the limit.

Since every physical space is created by substances (i.e., mass and energy), each physical space can be described by the speed of light.

The dimension of a physical space is determined by the speed of light, and the space is filled with substances (i.e., mass and energy).

Why light speed is limited?

It is because that our universe (a gigantic physical space) is filled with substances that cause the time delay of Electro-Magnetic (EM) waves propagation.

If there are physical substances that travel beyond the speed of light (and these remain to be discovered), then their velocities will also be limited, since our universe is compacted with substances.

In principle, substance can travel within a space without any time delay, if and only if the space is empty (or timeless).

However, empty space cannot exist in practice, since every physical space (e.g., our universe) must be fully filled with substances and (it has no empty space in it).

We note that, a timeless space in not a temporal space, and it does not exist within our temporal space.

Dark Energy & Dark Matters

It is known that our current universe is composed of 72 % Dark Energy, 23 % Dark Matters and the rest physical substances are about 5 %.

Since Dark Matters are in mass forms although they contributed about 23 % of the universal space, but they dominate the entire universal energy reservation over 95 percent, based on the Einstein's energy equation.

Big-Bang Theory

As we accepted the big bang theory for our universe creation, then the creation must start with Einstein's energy equation as given by the equation below:

$$\frac{\partial \varepsilon}{\partial t} = c^2 \frac{\partial m}{\partial t} = [\nabla \cdot S(v)] = -\frac{\partial}{\partial t}[\frac{1}{2} \in E^2(v) + \frac{1}{2}\mu H^2(v)]$$

where *M represents* a gigantic energy reservoir.

Based on the Big Bang Theory, it is apparently that, the creation was ignited by time and the exploded debris (i.e., matters and energy); they started to spread out in all directions like a balloon.

The boundary (i.e., radius of the sphere) of the universe expands at speed of light, as the created debris dispersed.

It took about 16 billion light years of chaotic evolution to come up with the present form of constellation, and our universe is still expanding at speed of light (beyond the current observation).

In view of recent Hubble space telescope observation, we see that galaxies at 15 billion light years and they are still moving away from us. This means that, the creation process has not stopped yet.

Due to intense convergent gravitational forces from the newly created debris of mass (e.g., galaxies, dark matters, and others), at the same time the universe might have started to de-create itself since the Big Bang started.

Trading time for space

Let us now take one of the simplest connections between temporal space (or physical space) and time, as given by distant equals to velocity times time.

$$d = v \cdot t$$

where d is the distance, v is the velocity and *t* are the time variable. This equation may be one the most unique representation in connecting between time and space!

A three-dimensional (Euclidean) physical space can be described by

$$(dx, dy, dz) = (vx, vy, vz) \cdot t$$

where (vx, vy, vz) are the velocity vectors.

Based on the current laws of science, speed of light is the ultimate limit. Then by replacing the velocity vectors with speed of light c, the Cartesian space coordinates can be described by,

$$(dx, dy, dz) = (c \cdot t, c \cdot t, c \cdot t)$$

Once again, we see that time can be traded for space and space cannot be traded for time since time is a forward variable. So, we see that, once time is expended, we cannot get it back.

Big Bang Creation

It is however more appropriate to describe the space creation by means of using spherical coordinate representation.

Based on the Big Bang theory, the boundary of our universe can be written as radius equals to velocity of light times time

$$r = c \cdot t$$

In which we see that the edge (or boundary) of our universe is expanding at the speed of light, as illustrated in figure 1.1.

Figure 1.1 shows that the creation of our universe starting from a big explosion, that is transforming from a dimensionless point-singularity into a three-dimensional space with time representation.

Our universe is an energy conservation dynamic bounded subspace.

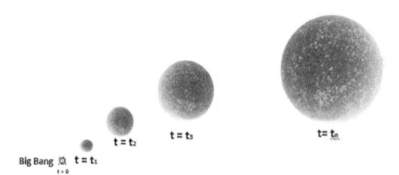

Figure 1.1

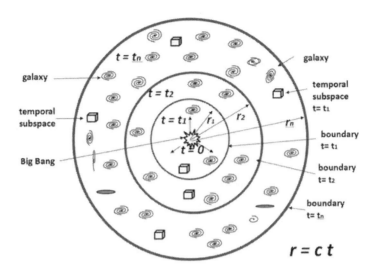

Figure 1.2 shows a composite temporal universe. We see that the boundary of the universe is limited by the speed of light times time (i.e., $r = c \cdot t$).

Conclusion

We have shown that time is an enigmatic variable in our universe. Without time there would be no matter, no energy no space and no life.

We have shown that energy and mass can be traded. We see that mass can be considered as an Energy Reservoir.

We have also shown that a physical space cannot be embedded in an absolute empty space and any physical space cannot have an absolute empty space within it. We see that every physical space must be fully packed with substances (i.e., energy and mass).

Since any physical space cannot be embedded in an absolute empty space, it is reasonably to see that our universe is a subspace within a more complex space, of course this remains to be found.

We have also shown that, it takes time to create a physical space, and the physical space cannot bring back the time that had been used for the creation.

Since all the physical spaces are created by time and substances (i.e., energy and mass), this implies the origin of our cosmos is created by time with a gigantic energy explosion.

We see that every substance must exist with time. Without time the ignition, the creation would not have taken place. Although time ignited the creation of our universe, it is the physical substances that show us the existence of time. We have concluded that our universe is a temporal space, and it is still expanding as based on current observation.

As we accepted the big bang theory of creation, we have seen that our universe has not reached to her half-life yet.

We are not alone is almost an absolute certainty. Someday we may find a right planet, once upon a time had harbored a civilization for a twinkle period of light time.

Remark

Essences of Temporal Universe (our home):
Science is a law of approximation and mathematics is an axiom of absolute certainty. Using exact math to evaluate inexact science cannot guarantee that the solution exists within our temporal universe.

"One important aspect of temporal universe is that one cannot get something from nothing; there is always a price to pay; every piece of temporal subspace (or every bit of information) takes energy and time to create.

Any science can be proven within our temporal universe is physically real; otherwise, it is fictitious unless it is validated by experiments."

The Burdens of Proof

Burden of a mathematical postulation:
- First, the burden is to prove that a solution exists.
- Second, the burden is to find the solution.

Burden of an emerging science:
- First burden is to prove its existence within our temporal universe.
- Second burden is to find out if we can afford to pay for it; in terms of energy and time.

(ΔE, Δt) or energy time intelligent information (ΔE, Δt, ΔI).

For example, creation of a simple facial tissue will take a huge amount of energy, time, and information (i.e., equivalent amount of entropy) to create the tissue. Let me emphasize that, "on this planet only humans can make it happen".

Stephen Hawking: A Genius

Professor Hawking was a world-renowned astrophysicist, a respected cosmic scientist, and a genius.

Although my discovery of temporal universe was started with the same root of big bang explosion, it is not a sub-space of Professor Hawking's universe. You may see from this presentation that my creation of temporal universe is somewhat different from Hawking's creation.

Nevertheless

The discovery of Temporal ($t > 0$) universe is the truest model we can use, based on our current knowledge of physics. As you will see in the following presentations, our universe is an Energy Conserved Dynamic Expanding Subspace that obeys all the laws of nature. I can summarize the properties of our universe with a simple equation:

$\Delta E\,(\Delta t)$

Where $\Delta E \approx \tfrac{1}{2} mc^2$

Δt changes from $\Delta t = 0$ to $\Delta t \to \infty$

CHAPTER 2

The Fate of Schrödinger's Cat

A peekaboo principle collapses by observation,

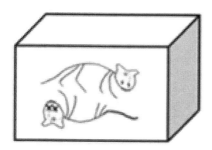 

yet, the fate of his cat cannot be decided in darkness

Paradox of Schrödinger's cat

Inside the box we equip a bottle of poison gas and a hammer to break the bottle. This in turn is triggered by the decaying of a radio-active particle, which in turn kills the cat.

Superposition is a fundamental principle of quantum mechanics. Superposition holds multi-quantum states of an atomic particle. In other words, the assumed two states radio-active particle can simultaneously coexist.

Since the hypothetical radio-active particle exist simultaneously in two possible quantum states (i.e., decay or non-decay), as imposed by the superposition principle of quantum mechanics, this means that the cat can be simultaneously alive and dead at the same time, before we open the box.

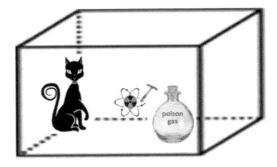

Figure 2.1: One of the laughable principles of all time, yet we have debated for decades!

But as soon we open the box the state of superposition of the particle collapses (w/o proof), in which we found the cat either alive or dead, but not at the same time.

This paradox in quantum mechanics has been intriguing quantum physicists for decades, since the birth of Schrödinger's cat in 1935 as he discloses it at a Copenhagen's forum.

With this, we are assuming to accept a fundamental principle of Quantum Mechanics which is that the principle itself was created in a Timeless Quantum Subspace.

However, a timeless subspace cannot exist within a temporal space, so that the solution as obtained by Schrödinger's equation contradicts the basic principle as a subspace within a temporal space.

Schrödinger's QM is Timeless

Because Schrödinger's quantum mechanics was built on an empty timeless subspace platform.

His fundamental principle only exists in a timeless subspace, and it does not exist w/in our temporal universe.

Was Schrödinger's Mistaken?

He introduces a timeless particle in the box, which is temporal.
But timeless and temporal cannot exist!
In the following image we show that Schrödinger's radio-active is timeless or time independent. But a timeless particle cannot exist within a temporal subspace. In which we see that the paradox of Schrödinger's cat, after all is not a paradox at all!

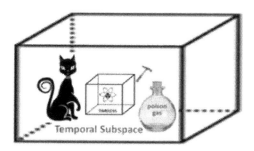

Figure 2.2 The paradox of Schrödinger's cat is leaving the cat alone we find that: "once upon a time, there was a half-live cat…"

Paradox of a half-boiled egg?

As we picked up a boiled egg, we cannot tell it is a soft-boiled or a hard-boiled unless we crack it open. But, if we keep it in the drawer for over a month, would you anticipate your boiled egg will be a hard boiled? If you knew the answer, then you will not take quantum computing seriously.

Irony of a great nearsighted physicist

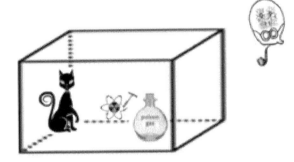

Figure 2.3: His superposition principle must wait until he touches the cat to collapse!

CHAPTER 3

What is "Wrong" with Theoretical Physicists

Time-space cannot be empty...

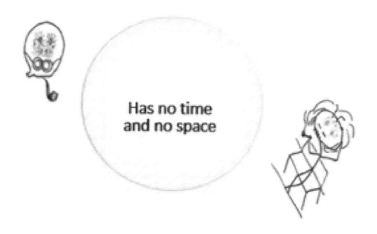

Yet, theoretical physicists have assumed it is possible for it to be empty for centuries!

Introduction to Physically Realizable Physics

What is "Wrong" with theoretical physicists?

A scientist's dilemma?

Should he/she bury within the timeless science and thrive, or change to temporal science in "light"?

As a learner, gradually we lose the "independent" logical thinking and opted to accept the approval of the others.

The essence of a "Price"

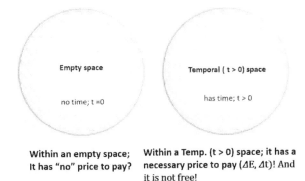

Within an empty space; It has "no" price to pay?

Within a Temp. (t > 0) space; it has a necessary price to pay ($\Delta E, \Delta t$)! And it is not free!

Figure 3.1

Temporal (t > 0) Universe

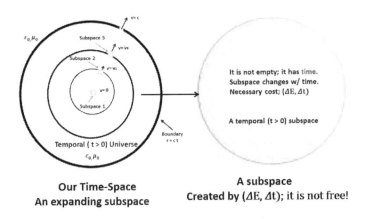

Our Time-Space
An expanding subspace

A subspace
Created by ($\Delta E, \Delta t$); it is not free!

Figure 3.2

Five Simple Elegant laws

1. Entropy: $S = -k \ln p$
2. Information: $I = \log_2 p$
3. Uncertainty; $\Delta E \, \Delta t \geq h$
4. Relativity: $\Delta t' = \Delta t / [1-(v/c)^2]^{1/2}$
5. Temporal Subspace: U: $\Delta E \, \Delta t \geq \Delta t \, \Delta mc^2$, $\Delta E \, \Delta t \geq h$

All these laws were developed from a timeless (t = 0) subspace, except temporal universe law.

All these laws are uncertainty principles since all of them change naturally with time. That is; w/o existence of time, we have "no" laws?

Nature of Δt

$\Delta I \approx \Delta E \cdot \Delta t = h$, per bit of information

$\Delta S \approx (\Delta E \cdot \Delta t)/T = h/T$, per bit of information

$\Delta E \, \Delta t \geq h$

$\Delta t' = \Delta t / [1-(v/c)^2]^{1/2}$

U: $\Delta E \, \Delta t \geq \Delta t \, (\Delta m) c^2$

Introduction to Physically Realizable Physics

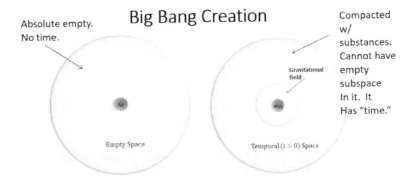

Big Bang Creation

Absolute empty. No time.

Empty Space

Not a physical realizable subspace: Emptiness and substance cannot coexist. Currently all cosmologists used this model. No Gravitational field no pressure, No Big Bang!

Compacted w/ substances. Cannot have empty subspace in it. It has "time."

Temporal (t > 0) Space

A physical realizable subspace: Over time; gravitational pressure to ignite the Big Bang? In which; BB more likely was originated. But gravitation force created by M. Yet not by time.

Figure 3.3

Relativity theory

Can be derived by Pythagoras Th.
$$\Delta t' = \Delta t / [1 - (v/c)^2]^{1/2}$$

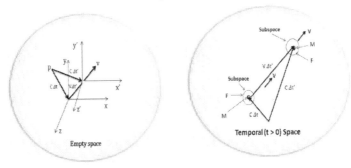

Empty space

Not a physical realizable subspace. Relativity is direction independent.

Temporal (t > 0) Space

A physical realizable subspace. Relativity is a directional principle.

Figure 3.4

Dilemma of a time traveler

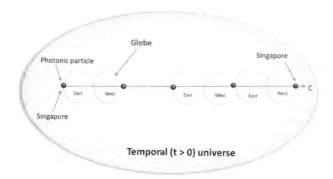

Figure 3.5 Since special theory is relativistic "directional"; the photonic traveler would "not" gain any time!

As John Wheeler said: "Space-time tells matter how to move; matter tells space-time how to curve". However, as I see it; it is time which tells space how to curve and vice versa.

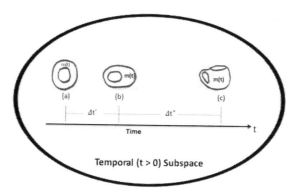

Figure 3.6 Time "curves" a subspace.

If we can move backward in time, then we see that, space curves the time? But we can't run backward in time! And this is precisely what the General Theory implies by treating time as an "independent" variable.

Schrödinger's Cat

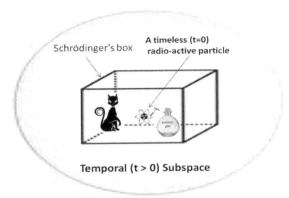

Figure 3.7

It is not a physical realizable hypothesis. Superposition principle is timeless, which is not existed w/in our universe!

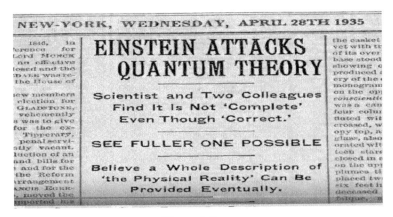

Figure 3.8

Superposition principle is "correct" but only within a timeless ($t = 0$) subspace. But it is "incorrect" within our temporal ($t > 0$) space".

General relativity theory

W/in Temporal subspace: time is coexisted w/ subspace, in which time is a dependent forward variable.

Q; can we curve space-time?

Ans; since our universe changes w/ time, it is not our universe changes the time?

Therefore, it is time-space curves with time, and it is "not" time-space curves the time since we "cannot" change the speed of time.

Timeless quantum entanglement

Pauli exclusive principle was developed on an "empty" space; it is a timeless principle!

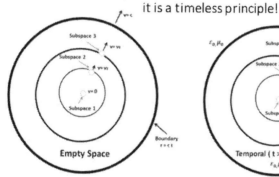 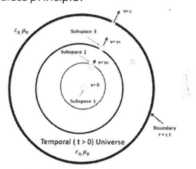

Universe has "no" time.
Entanglement is instant! But we are not living here?

Universe has time.
Entanglement is limited by speed of light.
We are living here?

Figure 3.9

Nature of Temporal (t > 0)

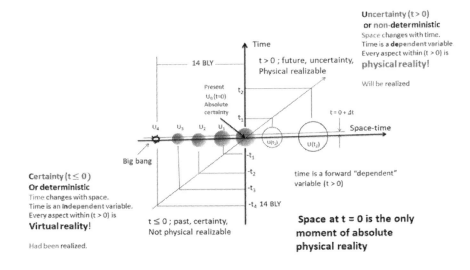

Figure 3.10

A necessary condition of physical reality

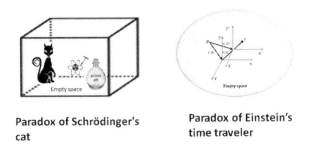

Figure 3.11 The tales of two-paradoxes: They are "not" physical realizable hypotheses?

Any hypothetical science is based of the past certainties to "predict" future outcome of uncertainties, which is supposed to be physical realizable. Otherwise, the solution will be virtual and fictitious as mathematics is. Therefore, imposition of physical realizable constraint [i.e., temporal (t > 0)] is a "necessary condition" to safeguard the solution would be physical realizable.

In which we see that; science is not supposed to be deterministic (i.e., approximation).

Why practically all the laws, principles and theories of science are timeless (t = 0)?

The root: Because all the laws, principles and theories were developed from a piece (or pieces) of empty paper for centuries, since theoretical physics is mathematics!

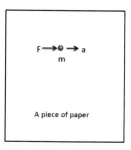

Figure 3.12

Conclusion

The notion of empty space contradicts physical reality.

A score of scientific hypotheses were virtual and fictitious.

Within our temporal universe, everything has a "cost" i.e., ΔE and Δt it is not free.

Remark:
- Science is a law of approximation.
- Mathematics is an axiom of certainty.

As a learner, gradually we lose the "independent" logical thinking and opted to accept the approval of the others.

References:

What Is "Wrong" with Current Theoretical Physicists ...

https://www.intechopen.com/online-first/what-is-wrong-with-current-theoretical-physicists-

Nature of Temporal (t > 0) Quantum theory: Part I

Nature of Temporal (t > 0) Quantum theory: Part II

CHAPTER 4

From Schrödinger's Equation to Quantum Conspiracy

We are accustomed to follow!

One of the greatest quantum conspirators?

Hamiltonian to Schrödinger Equation

Total energy of a Hamiltonian particle in motion is equal to its kinetic energy plus the particle's potential energy as given by,

$$H = p^2/(2m) + V$$

which is the well-known Hamiltonian operator, where p and m represent the particle's momentum and mass respectively and V is the particle's potential energy.

But it is a nonphysical realizable paradigm since substance and emptiness cannot coexist.

Figure 4.1

Hamiltonian Equation

$$H = - [h^2/(8\pi^2 m)] \nabla^2 + v$$

Where h is the Planck's constant, m and V are the mass and potential energy of the particle and ∇^2 is a Laplacian operator.

By virtue of energy conservation, Hamiltonian equation can be written as

$$H \psi = \{- [h^2/(8\pi^2 m)] \nabla^2 + V\} \psi = E \psi$$

Where ψ is the Hamiltonian wave function that remains to be determined, E is an energy factor which is to be assumed and V represents a potential energy that needs to be added in.

Schrödinger's equation

Since Schrödinger's quantum mechanics is the legacy of Hamiltonian, by adopting Bohr's quantum jump state energy E = h ν from Bohr's atomic model, Schrödinger's equation is given by,

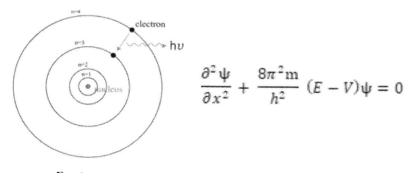

$$\frac{\partial^2 \psi}{\partial x^2} + \frac{8\pi^2 m}{h^2}(E-V)\psi = 0$$

Empty space

Figure 4.2

Where the nonphysical realizable Q. leap comes from?
It comes from Bohr's atomic model which was presented within an empty space as shown by the image above.

But my question is,

1. How can Q. leap radiate since Bohr's atom is embedded within an empty space?
2. But Bohr's atom is embedded within a temporal space, then Q. leap radiation must be temporal (t > 0). Which is a time and band limited wave function and exists only in the positive time domain (i.e., t > 0).

Schrödinger's equation,

$$\frac{\partial^2 \psi}{\partial x^2} + \frac{8\pi^2 m}{h^2}(E-V)\psi = 0$$

From which we see that it is a point-singularity deterministic time-independent, timeless (t = 0), or no-time equation. Which is virtually identical to Hamiltonian equation, except the energy factor; E = hv quantum state energy of Bohr.

The reason why this is a time-independent equation, firstly Schrödinger's equation is the legacy of Hamiltonian which is a timeless (t = 0) (i.e., time-independent) classical mechanics. But Hamiltonian was derived on an empty space platform, as has been shown in the figure 4.3.
Since Schrödinger's equation is a legacy of Hamiltonian, Schrödinger's equation is a timeless (t = 0) equation (i.e., as based of her empty space plat form).

Schrödinger's wave equation

Nevertheless, major differences between Schrödinger's mechanics and Hamiltonian must be the named sake of the Quantum, which comes Bohr's atomic quantum leap E = h υ, or a quanta of light that Schrödinger used for the development of his Q. mechanics. This precisely Schrödinger's solution is very similar to Hamiltonian.

$$\psi(t) = \psi_0 \exp[-i2\pi\upsilon(t-t_0)h]$$

Which is the wave equation where ψ_0 is an arbitrary constant, υ is the frequency of the quantum leap h υ and h is the Planck's constant.
But this is a nonphysical realizable equation, since it is not a time and band limited function and it is also not temporal (t > 0) (i, e., exists only within positive time domain).
Since all multi quantum leaps are not time limited, then it is trivial that all the Q. leaps will be instantaneously may not simultaneously superimposing at all time event within our temporal universe.

Physical Realizable Wavelet Function

However, it is possible to reconfigure a non-realizable wave function to be temporal (t > 0) as given by,

$$\psi(t) = \psi_0 \exp[-\alpha_0 (t - t_0)^2] \cos(2\pi \upsilon t), \; t > 0$$

From which we see that it is a time-limited temporal (t > 0) equation exists in t > 0, positive time domain that can be implemented within our temporal (t > 0) universe, as shown by.

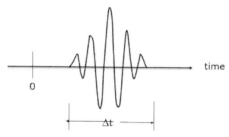

Figure 4.3

Schrödinger's Fundamental principle of superposition

More precise; every quantum state can be represented as a sum of two or more other distinct states. Mathematically, it refers to a property of solutions to the Schrödinger equation; since the Schrödinger equation is linear (Yet timeless t = 0), any linear combination of solutions will also be a solution.

Schrödinger's Superposition Principle: Multi-quantum leaps are simultaneously & instantaneously superimposing in space and in time.

What timeless (t = 0) space can do to particles?

Although timeless (t = 0) space is not a physical realizable paradigm; since emptiness and substance cannot coexist (i.e., Temporal Exclusive

Principle), but quantum physicists can implant particles into an empty space, since quantum scientists are also mathematicians.

Shows empty space can do to particles

Nevertheless, empty space has no time and no distance and no space from which we see that empty space is a virtual mathematical space, it is not a physical realizable space. And it cannot be an inaccessible space w/in our temporal (t > 0) universe.

Since empty space has no distance, we see that particle 1 and 2 are superimposing together over the entire empty space as shown on the righthand diagram. From which we see that this is precisely the

fundamental principle of superposition that Schrödinger stated. From which we see that Schrödinger's superposition principle only existed w/in a timeless (t = 0) empty space, which cannot exist w/in our universe.

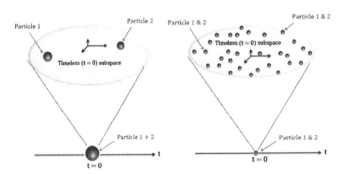

Figure 4.4

Time limited quantum state wavelets

Since there are time-limited temporal wavelets, which can be implemented within temporal (t > 0) universe. From which we see that these wavelets are not superimposing w/in our temporal (t > 0) universe as in contrast with the fundamental principle of superposition has predicted.

From which we have proven that Schrödinger's Superposition Principle is not existed w/in our universe.

$$\psi_{o1}(t) = \psi_{o1} \exp[-\alpha_{o1}(t-t_{o1})^2]\cos(2\pi\upsilon_{o1}t), \quad t > 0,$$
$$\psi_{o2}(t) = \psi_{o2} \exp[-\alpha_{o2}(t-t_{o2})^2]\cos(2\pi\upsilon_{o2}t), \quad t > 0,$$

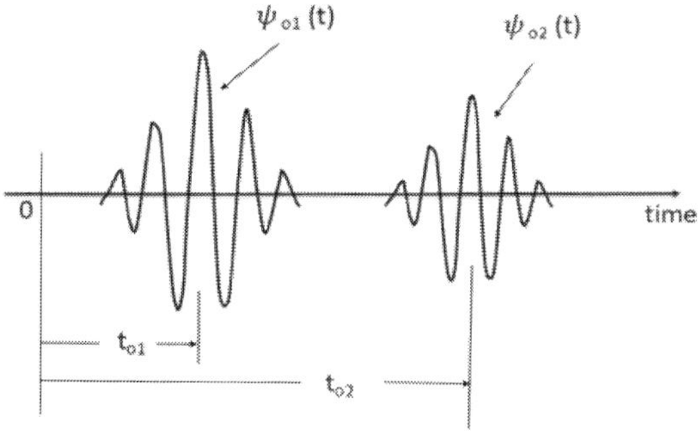

Figure 4.5

What empty space can do to quantum wavelets

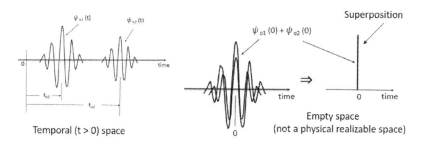

Figure 4.6

Since it is a timeless (t = 0) space, it has no time and no space. For which we see that; superposition exists within an empty space, and it does not exist w/in our temporal universe.

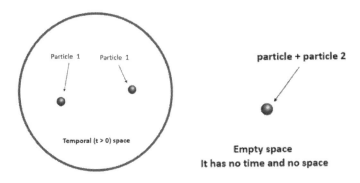

Figure 4.7

Figure 4.7 shows what happen to particles existed w/in our universe. For example, we assume particles 1 and 2 are plunged w/in a temporal space as shown in the figure. We see that there are precisely located at their specific locations within a temporal (t > 0) space, since the temporal space has time and distance. One condition I stress that; particle in motion or static is a temporal particle that changes naturally w/ time. In which we see that Schrödinger's superposition principle fails to exist within our temporal (t > 0) universe.

It is interesting to see what happen if this particle is plunged w/in an empty space, although it is not a physically realizable hypothesis as depicted on the right-hand side. From which we see that particles lost their position identities by superposing at time t = 0.

Once again, I have proven that superposition principle exists only w/in a mathematical virtual timeless (t = 0) space, but not w/in our universe.

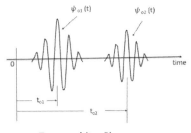

Figure 4.8

Figure 4.8 shows two time-limited multi-quantum wavelets exist w/in a temporal (t > 0) space. From which we see that superposition fails to exist w/in our temporal (t > 0) universe, since w/in temporal universe it has time and space.

Let me stress that, within our universe everything has a price to pay; a section of time Δt and amount of energy Δt or space. But it is the section of Δt that we cannot squeeze to zero (i.e., Δt = 0,) but it can be approached

to zero (i.e., Δt → 0). This is causality or temporal (t > 0) constraint of our universe.

Temporal (t > 0) Universe

Figure 4.9 shows our universe started from big bang creation changes naturally with time, it shows the age of our universe is about 14 billion light years (BLY) old.

Past time domain (t < 0) is the certainty virtual images without physical substance and time.

Future time domain (t > 0) represents a physical realizable uncertainty universe.

While t = 0 represents an instant moment of absolute physical reality of our universe.

Past universes have changed with time; time can be treated as an independent variable.

Future universe changes with time; time is a dependent variable.

Present universe (t = 0) is the absolute moment of physical reality

Therefore, it is a mistake to treat our temporal (t > 0) universe as a four-dimensional space-time continuum as most scientists do since time & universe coexist.

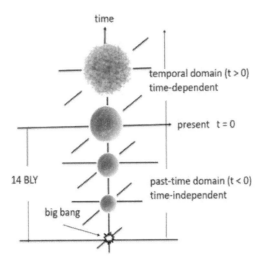

Figure 4.9

Conclusion

Since Quantum Supremacy depends on Schrödinger's fundamental principle of superposition, as I see it; the timeless principle has emerged as a worldwide Quantum Conspiracy!

CHAPTER 5

The Limits of Einstein's Theory of Relativity

Paradox of atomic clocks

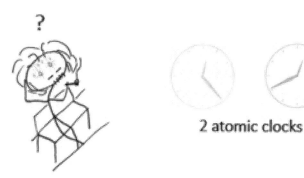

Either clocks are right, or physical reality is correct?

The limits to Einstein's Theories pf Relativity

Questions all scientists face:
1. Should I uncritically accept the presuppositions on which today's theories rest—especially when doing so will ensure me greater acceptance in the scientific world?
2. Should I pursue only that which will further my career, even if doing so may prevent me from discovering more fundamental truths?

Generalizations:
1. Even the most independent thinkers will, over time, tend to become defensive of their own theories, and consequently tend to overlook possible flaws in and exceptions to them.
2. Most people will defend orthodoxy even in the face of incontrovertible evidence.
3. Mathematics tends to legitimize any theory in both the academic world and the public mind, sometimes making the work seem more important, more serious, truer, and more real than it really is.

Some of the fundamental differences between science and mathematics

"Ontology is a prerequisite for physics, but not for mathematics. It means physics is ultimately concerned with descriptions of the real world, while mathematics is concerned with abstract patterns, even beyond the real world. Thus, physics statements are synthetic, while mathematical statements are analytic. Mathematics contains hypotheses, while physics contains theories. Mathematical statements have to be only logically true, while physics statements must match observed and experimental data. The distinction is clear-cut, but not always obvious." –Saheed Veradi, Sahand University of Technology, Tabriz, Iran.

"Mathematics, the science of structure, order, and relation that has evolved from elemental practices of counting, measuring, and describing the shapes of objects. It deals with logical reasoning and quantitative calculation, and its development has involved an increasing degree of idealization and abstraction of its subject matter."

-Fraser, Craig C. et. al. *Encyclopedia Britannica "Mathematics."* November 09, 2020. https://www.britannica.com/science/mathematics (accessed November 22, 2020).

Because physics often relies on mathematics, one could go as far as to say that physics needs mathematics. The converse, however, is not necessarily true; mathematics does not necessarily need physics.

Simply because physics utilizes and often relies on mathematics does not mean that physics and mathematics are equivalent. Mathematics can be thought of as a tool—one, incidentally, that is often used in physics as well as a host of other fields.

In its most abstract forms, mathematics involves "an increasing degree of idealization." As a result, and similar to language, certain forms of abstract mathematics do not have to correspond to physical reality.

This is why certain mathematical principles may be no more a reflection of how the world actually works than particular principles, theories, or ideas in economics, political theory, or philosophy.

Unlike in mathematics, every law, principle, theory, postulate, or paradox in the realm of physics must comply with physical reality. In other words, each must comply with the actual conditions of our universe where the passage of time is unvaried and never stops and never goes backward.

As a result, I am defining our universe as a temporal (i.e. time-space) universe where $t > 0$.

- Any scientific law, principle, theory, postulate, or paradox that does not comply with physical reality, cannot exist in a temporal universe.

- Einstein's special and general theories of relativity are no exception.

Empty space paradigm

Einstein's special theory of relativity can be derived from Pythagoras' theorem within an empty space as given by.

$$(c \Delta t')^2 = (c \Delta t)^2 + (v \Delta t')^2$$

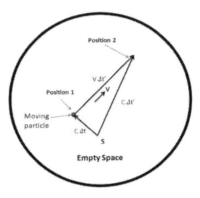

Figure 5.1

Thus, $\Delta t' = \Delta t / [1 - (v/c)^2]^{1/2}$ where $\Delta t'$ and Δt are the relativistic time dilations of the moving and static subspaces respectively, v is the velocity of a moving subspace, c is the velocity of light and S is the light emitter. Therefore, the Special Theory is relativistically directionally independent.

Temporal (t > 0) space paradigm

If the derivation starts at time $t = t_1$ from light source S, it will take time Δt (i.e., $t = t_1 + \Delta t$) for light particle 1 to reach position 1. Δt is a subsection of the space-time continuum for light particle 2.

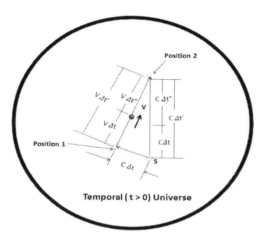

Figure 5.2

Light particle 2 reaches position 2 after light particle 1 reaches position 1. In other words, since (v·Δt) is a sub-distance of (v·Δt'), light particle 1 reaches position 1 before light particle 2 reaches position 2. Thus Δt < Δt', meaning that light particle 2 reaches position 2 after light particle 1 reaches position 1. In other words, since (v·Δt) is a sub-distance of (v·Δt'), light particle 1 reaches position 1 before light particle 2 reaches position 2. So, it will take light particle 2 an additional time of Δt" to reach position 2.

Because light particle 2 must travel an extra distance of C Δt" = C (Δt' − Δt) to reach position 2, it must reach its destination sometime after light particle 1 reaches position 1.

As I see it, the moving particle has no section of time-gain relative to the static position 1, since time at position 1 and 2 are "the same" (t = t$_2$ = t$_1$ + Δt') when moving particle reaches position 2. The time elapsed at position 1 is actually Δt' = Δt + Δt", instead of Δt as it would be according to the special theory of relativity.

From figure 5.2 we see that the time elapsed by the time particle 1 reaches position 1 is actually Δt' = Δt + Δt", instead of just Δt as assumed by the special theory of relativity.

$$\Delta t' = \Delta t / [1 - (v/c)^2]^{1/2}$$

Given the above, under what conditions is the special theory of relativity legitimate within our temporal (t > 0) universe? (Note that by raising this question I am not discounting the value that Einstein's theory has in producing one of the important equations in modern physics (i.e. E = mc², where E is energy, m is mass and c, the velocity of light, demonstrating that energy is equivalent to mass.).

The special theory was derived under the assumption that mass is at rest and therefore, not in motion. But from the standpoint of physical reality, mass cannot be totally at rest because it is a temporal (t > 0) substance and therefore, constantly changing with time.

Special theory fails to exist?

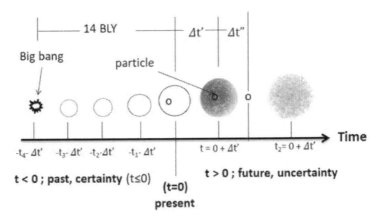

Figure 5.3: Time-Space diagram

Figure 5.3 shows the age of our universe is about 14 billion light years old. The past time domain (t < 0) corresponds to a certain past, or what I am defining as virtual events.

The future time domain (t > 0), on the other hand, corresponds to uncertain physical consequences. The present moment at t = 0 continuously progresses to a new present moment according to the equation,

$$t = 0 + \Delta t'$$

The present moment corresponds to an absolute certain moment. It is impossible for a particle, no matter how small, to travel ahead or behind the pace of time-space within our universe, if it is to be part of our real and temporal universe rather than part of a fictional universe.

Paradox of a time traveler

Einstein's special theory of relativity tells us we can experience a slowing down of time relative to someone if we are traveling faster than they are. But the rules of a temporal universe show us this is impossible; we cannot change time. We can only change with time. Space changes with time. It does not, however, change time.

Nature of Temporal (t > 0)

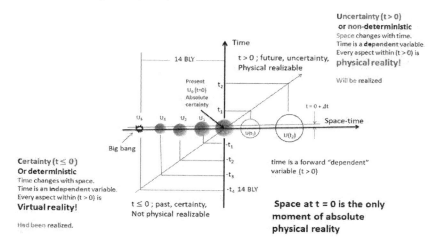

Figure 5.4

Necessary Conditions

Since present moment (t = 0) divides physical and virtual realities. Time must be greater than zero (i.e. t > 0) if any solution is to be physically realizable, or in other words, based on reality rather than fiction.

Anything considered real science should be physically realizable. If not, it should be considered as a pseudoscience.

Science is not deterministic. In other words, science can only approximate reality with either more or less accuracy, with either more or less precision, and from a greater or lesser number of perspectives; it can never fully determine reality.

Paradox of classical and modern science

Since the general theory of relativity is based on a time-independent deterministic variable, it is a deterministic theory instead of a time-dependent non-deterministic theory.

As a result, the general theory of relativity is not physically realizable within our temporal (t > 0) universe; space curves with time, but space does not itself curve time.

Note that practically all of classical science has used past deterministic certainties to analyze future outcomes.

This is precisely why all scientific laws, principles, theories, and paradoxes are fixed and deterministic, rather than changing naturally with time. (And this includes some of the probabilistic distributions such as wave function, Boltzmann distribution and many others, that do not change with time.)

Nature of Time

Once again, it is the present moment, where t = 0, that divides physical and virtual realities. Only in the present moment (t = 0) can we have absolute certainty in the material world. In other words, only the present moment (t = 0) corresponds to physical reality.

In contrast, the past (t < 0) corresponds to virtual and thus, certain consequences, while the future (t > 0) corresponds to physical and thus, uncertain consequences.

In short, the universe and its subspaces only exist in the absolute material sense, in the present moment, when t = 0.

All past consequences are virtual and certain events but with no physical substance or time dependency. All past consequences therefore constitute information.

Similarly, all future consequences are physically realizable and change with time.

Non-realizable General Theory

According to John Wheeler, "Space-time tells matter how to move; matter tells space-time how to curve." In contrast, I am proposing that time tells space how to curve while space does not tell time how to curve.

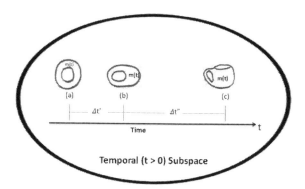

Figure 5.5

Figure 5.5 demonstrates how time "curves" a subspace. If we could reverse time, we might be able say that space curves time. The problem is that we cannot reverse time.

Yet, the possibility of reversing time is precisely what the general theory of relativity is proposing when it treats time as an independent variable.

In conclusion

The special theory of relativity does not apply to our temporal ($t > 0$) universe, because the theory presupposes the possibility of empty subspace.

CHAPTER 6

Quantum Qubit Information Conspiracy

Timeless qubit conspiracy

 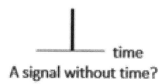

A signal without time?

Is information moved by time, or information-transmission without time?

A Conventional Communication Channel

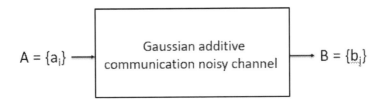

Figure 6.1

Information-transmission through an additive Gaussian noisy channel is shown in figure 6.1 where A = {a_i} and B = {b_j} are the input and output ensembles respectively.

Quantum Information-transmission

Two key mutual information equations through an additive Gaussian noisy channel are given respectively by,

$I(A;B) = H(A) - H(A/B)$
$I(A;B) = H(B) - H(B/A)$

where $H(A)$ represents an information (i.e., entropy) provided by the sender, $H(A/B)$ is the information lost (or equivocation) through the channel due to noise, $H(B)$ is the amount of information received at the receiving end, and $H(B/A)$ is the noise entropy of the channel.

Noiseless Communication Channel

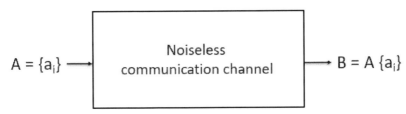

Figure 6.2

```
A = {aᵢ} ─────────────────→ B = {aᵢ}
H(A)                          H(B)
```

$I(A;B) = H(A) - H(A/B) = H(A)$
Since $H(A/B) = 0$. (i.e., no ambiguity)
$I(A;B) = H(B) - H(B/A) = H(A)$
Where $H(B/A) = 0$; (i.e., no noise). Since $H(B) \geq H(A)$; $H(B) = H(A)$.

However, there is a distinction between these mutual information equations; one is for reliable-transmission and the other is for information-retrieval.

Nevertheless, there are two kinds of communication orientations:
- One is for virtuous information-retrieval of Norbert Wiener,
- and the other is for reliable transmission of Claude Shannon.

Wiener's communication strategy is that; to recovered information that has been corrupted through transmission, but it has a "heavy price" to pay for post processing.

Shannon's information-transmission carries a step further by encoding the information before it is transmitted such that, information can be "reliably" received, which also has a "price" to pay, but mostly at the transmitting end.

Nevertheless, the distinction is Weiner type is to retrieve the information after it has been received, and Shannon type is to protect the information before it is sent.

From this, we see that reliable information transmission is to minimize the noise entropy H of A over B (or equivocation), as shown by. $I(A; B) \approx H(A)$

Nevertheless, a simple way to achieve is increasing the signal energy (i.e., ΔE) to noise ratio. And other is by redundancy coding.

On the other hand, information retrieval is to maximize the noise entropy $H(B/A)$ (i.e., the channel noise), such that mutual information $I(A; B)$ approaches to input entropy $H(A)$ as given by,

$$I(A; B) = H(B) - H(B/A) \approx H(A)$$

Since $H(B) \geq H(A)$, we see that if and only if $H(B/A) = 0$, which is equivalent to a lossless or noise free channel.

In view of quantum qubit information-transmission, which is basically using Wiener's communication strategy, however a much higher price to paid at the receiving end (e.g., longer section of time Δt is needed) and much more costly (i.e., larger energy ΔE required). But moves further away from instantaneously transmission or for computing.

Since quantum qubit information-transmission requires the receiving end entropy $H(B)$ be "more equivocal" (i.e., uncertain). We see that it treats $H(B)$ as a source entropy of $H(A)$.

But $H(B) \geq H(A)$, it does not mean that increases entropy $H(B)$ at the receiving end changes source entropy $H(A)$ at the transmitting end, since $H(B)$ can never be equals to $H(A)$, unless it is w/in a noiseless information channel.

Nevertheless, noiseless paradigm is not a physically realizable paradigm since every transmission change with time. In other words, noiseless paradigm is virtual as mathematics is, which cannot exist w/in our temporal ($t > 0$) universe.

Quantum qubit information channel

Since input information source of $A = \{0, 1\}$ is a binary digital source, we see that input entropy $H(A) = 1$ bit per Δt.

Why a section of time Δt is needed because every bit needs a section of Δt to represent a time-signal, no matter how small Δt is, but it cannot be equaled to zero (i.e., $\Delta t = 0$).

And each section of Δt is a part of $(\Delta t, \Delta E)$, which is the necessary cost.

Nevertheless, w/in an empty space it has no time, my question is that how can we transmit a time-signal without a section of Δt? which we cannot.

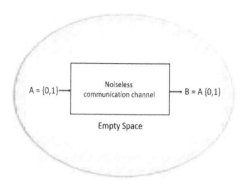

Figure 6.3

In view of this paradigm, we see that it is not a physically realizable paradigm which is because empty space is not an accessible subspace w/in our temporal universe, by means of temporal exclusive principle.

Since every bit of information can be transmitted instantly within an empty space which has no time and no energy to pay, then input information source A is capable to transmitting a binary signal of 0 and 1 instantaneously and simultaneously. From which we see that the instantaneously and simultaneously transmission of 0 and 1 is precisely what Schrödinger's fundamental principle of superposition has stated. This is the reason that output H(B) had been treated as a qubit entropy by quantum scientists as given by,

- $H(B) = - \log_2 p(qubit)$, bits
- $H(B) = - \log_{qubit} p(qubit)$, qubits

A typical noiseless communication channel

For example, if one sent a 0 or 1 message within an envelope to you, you would not know until you open the envelope, although you know it is a binary yes and no message.

But it is "not" your consciousness changes the outcome of the sent message. Similarly, before you crack open a boiled egg you do not know if the boiled egg is hard or soft boiled, even if you know it has been boiled.

Guessing correctly which one it is does not mean your consciousness divined the outcome.

Legitimacy of Quantum qubit information channel?

From which as I see it, quantum information-transmission is basically designed to extracting information as Weiner's type communication strategy, which requires post processing but with a bigger price to be paid.

Since it needs post processing, transmitted signal deviates further away from instantaneous (i.e., t = 0) transmission merit, even though we assume qubit transmission works. Nevertheless, instantaneous (i.e., t = 0) transmission is not a physically realizable transmission. From which we see that qubit information is not a legitimized physically realizable information, as from temporal exclusive principle standpoint.

Can qubit information formulate w/in our time-space?

Since within our temporal (t > 0) universe, everything has a price to pay; a section of time Δt and an amount of energy ΔE, for which every qubit needs a section of time Δt to transmit no matter how small it is (i.e., $\Delta t \rightarrow 0$) but never equals to zero ($\Delta t = 0$), otherwise a qubit information cannot be represented as a time-signal w/in our universe.

In words, w/o a section of time Δt to present a qubit information, there will be no qubit physical reality can be transmitted w/in our time-space. And as well w/in an empty space since empty space has no time and on space.

Figure 6.4 shows a rectangular pulse time-signal on the left and a wavelet time-signal on the right. From which we see that every signal needs a section of time Δt and an amount of energy ΔE (i.e., $\Delta t, \Delta E$) to represent a temporal signal.

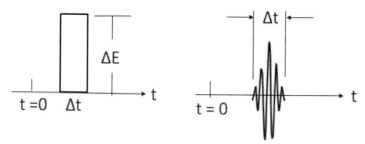

Figure 6.4

Nevertheless, aside severity of nonexistent superposition principle, it must be the time-signal representation w/in our temporal (t > 0) space. Since quantum qubit communication can only exist w/in an empty virtual space, it cannot physically realize w/in our time-space. From which we see that qubit information is virtual as mathematics is, which cannot be implemented w/in our temporal (t > 0) universe.

Finally let me hypothesize a noiseless communication scenario as shown in figure 6.5:

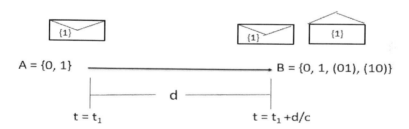

Figure 6.5

In which we assume a "1" message was inserted w/in an envelope sent through a noiseless (i.e., w/out contamination) channel to a receiver at B. Since receiver does not know the message w/in the envelope until he opens it. Which is similar to the half-life cat of Schrödinger, until the observer opens Schrödinger's box. From which we see that, the message w/in the envelope had been determined before it was sent, but it is not the receiver's.

consciousness changes the outcome of the message. Like the life of Schrödinger's cat which had been determined before the observer opens Schrödinger's box. Moreover, we see that transmission is not instant, since every bit takes a section of time Δt to transmit and also an amount of energy ΔE to make it happen, and it is not free.

Conclusion

Firstly, we had shown qubit information-transmission depends on Schrödinger's fundamental principle of superposition, but the principle is not actually existed w/in our temporal ($t > 0$) universe.

Secondly, we had shown that qubit quantum information is virtual as mathematics, it has no section of time to formulate a time-signal. In other words, qubit information only exists within an empty timeless ($t = 0$) space where clock is not ticking, but it is not existed within our universe where clock ticks!

From all of this we can see that qubit information is as elusive as Schrödinger's cat, and it is hard to see that qubit information had become a worldwide Qubit Conspiracy, which is fictitious as Schrödinger's half-life Cat!

CHAPTER 7

Why modern physics is so weird... and so wrong

So weird we deserve not to understand it?

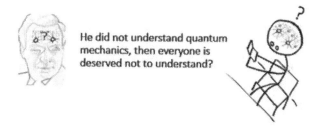

He did not understand quantum mechanics, then everyone is deserved not to understand?

Theoretical physicists know how a particle travels from point A to point B, but they do not know why.

A few preliminary remarks

Science is meant to discover but not to create the laws of nature.

Science is a way of describing physical reality. As a result, it should correspond to physical reality and should not be deterministic.

Given the proper evidence and logic, scientific laws, principles, and theories can always be revised, changed, and even refuted.

Everything within our universe has a necessary price to pay in the form of time Δt and an amount of energy ΔE. Nothing exists in timeless space, and nothing exists without an expenditure of energy. Nothing, in other words, comes for "free."

Within the paradigm of quantum physics, things change at the speed of light. The problem with quantum physics is that it assumes we can change time.

Modern Physics versus Physical Reality

Einstein's theory tells us we can change time. Yet, we can't.

Schrödinger's principle tells us one thing can exist in two places at the same time. Yet, they can't.

Heisenberg's uncertainty is independent of time. Yet, everything is time dependent.

Bohr's quantum leap is not time limited. Yet, everything is time limited.

Pauli's entanglement is not space limited. Yet, everything is space limited.

Dirac posits that anti-particles exist within our universe. Yet, anti-particles are independent of time and therefore cannot exist.

Feynman's QED and Hawking tell us that black holes exist within our universe. Yet, this is a form of science fiction conjured up by abstract mathematics.

Why is modern physics so wrong?

Practically all the theories, principles and laws in modern physics have been derived from a 4-d spacetime continuum of Einstein.

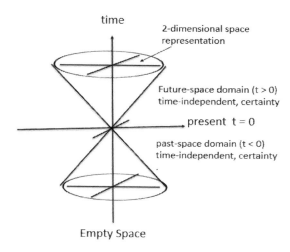

Figure 7.1

Modern physics posits a zero-summed subspace that allows anti-matter to exist.

Time is treated as an independent variable even though doing so violates the 2nd Law of Thermodynamics. This paradigm does not reflect physical reality.

Mathematics is not science.

Without mathematics, science would be extremely limited. This is because science often uses mathematics and mathematics can help scientists make new discoveries. Yet, mathematics is not equal to science. It is simply a tool frequently used by scientists. Unfortunately, many people, including scientists and mathematicians themselves, have conflated the two fields, acting as if everything mathematical has a physical, real-world representation. As a result, science has for decades been hijacked by mathematics.

Mathematics, no matter how rigorous, does not correspond to physical reality. Only mathematics that is developed from a physically realizable platform can correspond to physical reality.

What is a physically realizable platform? It is the temporal (t > 0) subspace that we are living in, where every subspace changes with time. This means that even the most elementary particles in the universe exist in temporal subspace.

Since our universe is a time-dependent subspace where time is a forward variable running at a constant pace, it is impossible to change or stop time. And this is precisely one of the fundamental laws of nature: everything changes with time. An example:

Take the mass-energy equivalent equation:

$$E = ½ mc^2$$

If written as a zero-summed equation, it becomes one of the two following:

$$E - ½ mc^2 = 0, \quad 0 = ½ mc^2 - E$$

Mathematically speaking these equations are the same. They demonstrate in the purely mathematical, that negative energy and anti-mass can exist. In temporal subspace, however, this is not true. Neither negative energy nor anti-mass can exist. For example, take the Dirac equation:

$$i \partial \psi - m\psi = 0$$

Which hypothesizes that anti-matter and negative energy can exist.

Yet, if Dirac equation is written as the following non-zero summed form,

$$i\partial\psi = m\psi$$

he might not have had postulated the existence of anti-matter and negative energy.

Once again, mathematics is not equivalent to science.

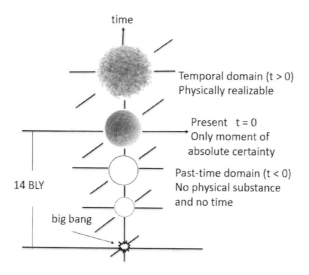

Figure 7.2

A physically realizable temporal universe is a non-zero summed, energy conservation subspace. Physical reality only exits in the present (t = 0) moment. Past time (t < 0) domain represents certain memories which have no physical manifestation other than synapses in our brain. The future (t > 0) represents a physically realizable universe that is uncertain. Thus, science can be thought of as a means of approximating rather than determining the future.

A classic example:

A classic text-book example to derive Einstein's special theory. In figure 7.3, the clock within the moving set of reflected mirrors is ticking more slowly than the stationary clock.

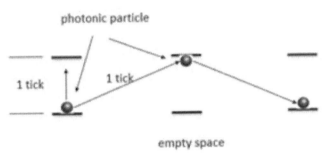

Figure 7.3

In contrast, within a temporal (t > 0) subspace, the clock within the moving platform does not slow down.

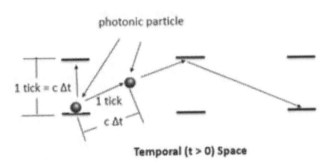

Figure 7.4

In our universe, God does play dice and Schrödinger's cat cannot exist in two places at one time.

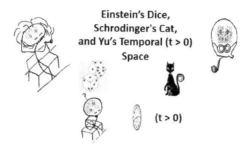

Figure 7.5 The Art of a Physically Realizable Principle

What does the previous image indicate?

When Einstein said, "God does not play dice," I believe he meant science should be able to make predictions with absolute certainty. In reality, however, science is merely a form of approximating physical reality.

On the other hand, Schrödinger's cat indicates that predictions are never absolutely uncertain. Yet in reality, scientific predictions should not be totally ambiguous.

Temporal (t > 0) space tells us that every prediction changes with time. In other words, the further away we travel from current absolute certainty the more uncertain our prediction. Science, once again, is a means of approximating physical reality.

Finally, within our universe we cannot get something from nothing; there is always a price to pay. This means that every physical phenomenon costs a section of time Δt and an amount of energy ΔE.

30 billon dollars has so far been committed to the global quantum computing initiative! And this is just one of many global research initiatives in 2022, including the search for exotic particles, wormholes,

and black hole. Why are we committing so much money to the realm of science fiction rather than science?

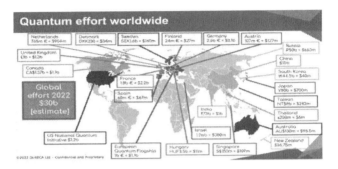

Figure 7.6

Please check out my newly published book:

Figure 7.7

CHAPTER 8

Does Einstein's General Theory Belong to the Realm of Science?

Can you visit another version of yourself?

Believe it or not, time traveling in planet of apes is a fictitious story?

Special relativity is fictional!

Einstein theorized in one of his famous thought experiments that light should bend as a result of gravitational force.

Using the equation, F = ma, from Newton's second law of motion, Einstein demonstrated that gravitational force is proportional to acceleration.

Because light entering an accelerating spaceship appears to bend to an observer inside the spaceship, Einstein theorized that gravity exerts a "pull" on light just as it does on matter. Light, however, is not matter. It is an electromagnetic wave and is not affected by gravity.

In the following text, I will demonstrate that gravitational force does not bend light.

Because Einstein's general theory of relativity is rooted in the false assumption that gravity bends light, the theory is itself faulty.

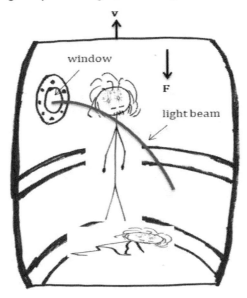

Figure 8.1

In Figure 8.1 Einstein is seen floating inside a spacecraft that is itself, floating in outer space. If the spacecraft accelerates upward with sufficient force (F), Einstein will be pulled downward to the floor, with the acceleration acting as gravity would if he were standing on the earth's surface. If a light beam enters through the window, it too, will travel in a downward arc relative to Einstein. As a result, the acceleration appears to be pulling the light beam downward as it travels.

Einstein formulated his light bending hypothesis by imagining the path of a light beam as it entered the window of an accelerating spaceship. If the spaceship were moving close to the speed of light when the light beam entered, the light beam would theoretically travel in an arc relative to the person in the spaceship. Because the acceleration of the spaceship was at the same time generating a force that pulled the person to the floor of the spaceship, thereby enacting a pseudo-gravitational force within the spaceship, Einstein hypothesized that gravity and acceleration possessed the same properties. If so, he deduced, then gravity had the power to "bend" light.

Einstein formulated his light bending hypothesis by imagining the path of a light beam as it entered the window of an accelerating spaceship. If the spaceship were moving close to the speed of light when the light beam entered, the light beam would theoretically travel in an arc relative to the person in the spaceship. Because the acceleration of the spaceship was at the same time generating a force that pulled the person to the floor of the spaceship, thereby enacting a pseudo-gravitational force within the spaceship, Einstein hypothesized that gravity and acceleration possessed the same properties. If so, he deduced, then gravity had the power to "bend" light.

In figure 8.2, the person in the spaceship sees a light beam entering the window and "bending" until it reaches a lower level inside of the spaceship. Einstein hypothesized that gravity works in the same fashion and therefore bends light. (Note that v is the velocity of the spacecraft and F is the gravitational force.)

Introduction to Physically Realizable Physics

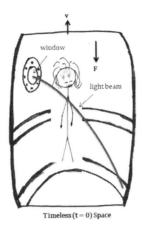

Figure 8.2

From Newtonian Mechanics to Einstein's Relativistic Mechanics

According to Newton's second law and his gravitational law, force (F) can be represented as follows,

$$F = mg \text{ and } F = (G\, m\, M)\, r^2$$

where m and M are the masses of a given particle and planet earth respectively, g is the gravitational acceleration, G is the gravitational constant and r is the distance. Combining the two equations, we have,

$$g = (G\, M)\, r^2$$

From this we can see the profound connection between mass and gravity.

Einstein based his general theory of relativity on the assumption that light is made up of particles that possess wave-like properties. Since light is a transversal electromagnetic wave, however, it has no mass and therefore cannot interact with gravity. Einstein treated time as an independent variable. His hypothesis assumed a timeless (t = 0) empty subspace, even though the principle of temporal exclusivity means that timelessness cannot exist within our temporal (t > 0) universe.

Empty timeless (t =0) space

Figure 8.3

Einstein's light bending hypothesis assumes the possibility of timelessness (t = 0). A light beam enters the window of an accelerating spaceship from point A. Because of the acceleration, the light beam strikes the opposite side of the cabin at point B, which is lower than point A. Because the spaceship's acceleration has pulled the person to the floor, acting therefore, as a pseudo-gravitational force, Einstein theorized that gravity and acceleration were one and the same. In doing so, he posited that gravity "pulls" or "bends" light.

Introduction to Physically Realizable Physics

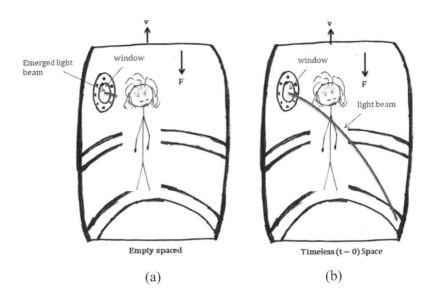

Figure 8.4 (a) shows a light beam enters window at time t = t', and point A, (b) shows that the light beam is observed at a lower point in the spaceship at time t = t".

In figure 8.4(a), the person in the accelerating spaceship sees a light beam entering the window from point A, at time t = t'. In figure 8.4 (b), the light beam strikes a lower point on the opposite side of the moving spaceship at a later time, t = t".
Einstein's light bending hypothesis turns out to be false in a temporal (t > 0) subspace. Just as it is outside of the spaceship, time is necessarily a dependent variable inside of the spaceship. In fact, everything existing inside the spaceship continues to exist within our temporal (t > 0) universe. Everything, whether inside or outside of the spaceship therefore, changes with time.

Einstein's light bending hypothesis turns out to be false in a temporal (t > 0) subspace. Just as it is outside of the spaceship, time is necessarily a dependent variable inside of the spaceship. In fact, everything existing inside the spaceship continues to exist within our temporal (t > 0) universe. Everything, whether inside or outside of the spaceship therefore, changes with time.

In our temporal universe, time ticks at the same rate everywhere, including both inside and outside of the spaceship. If the distance from A to B' is longer than from A to B, it means that according to Einstein's hypothesis, either light must travel faster than its given constant (c), once it enters the spaceship, or time must slow down.

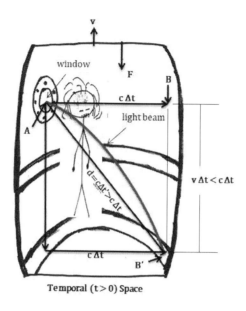

Figure 8.5

Since light is independent of gravity in our temporal (t > 0) universe, the path of the light beam should not change once it enters the spaceship. The light beam strikes point B' is actually due the upward motion of the spaceship, rather than any purported effect of gravity. The misguided conflation of acceleration with gravitational force has resulted in scientists generating fictional scenarios, or in other words, theories that should be relegated to the realm of science fiction. Stephen Hawking's theories on black holes and wormholes are but two examples of science fiction being paraded as science.

Introduction to Physically Realizable Physics

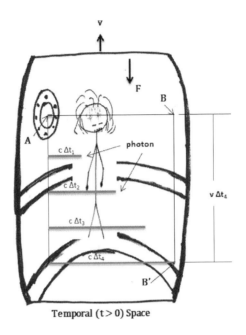

Figure 8.6

Figure 8.6 shows a time-composited diagram of Einstein's hypothesis. The light beam (i.e., photon) moves across the spaceship at the speed of light and is unaffected by gravitational forces. The path of the light beam consequently does not change even though it appears to travel in an arc relative to someone inside the moving spaceship. Each red line represents the path the light beam has traveled relative to a fixed point outside the spaceship after time Δt_n has elapsed. The endpoint of each line represents where inside of the spaceship the light beam will end up after time Δt_n has elapsed, where v is the velocity of the spaceship, $\Delta t_1 < \Delta t_2 < \Delta t_3 < \Delta t_4$ (each of which is a section of time) and F is the gravitational force.

Introduction to Physically Realizable Physics

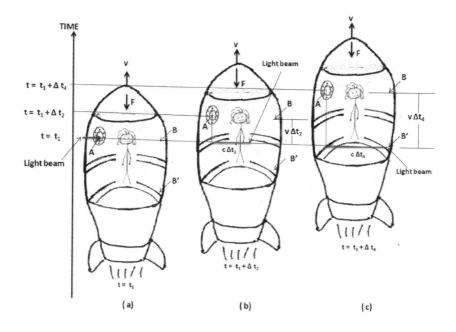

Figure 8.7 This figure provides a different perspective on the phenomenon described.

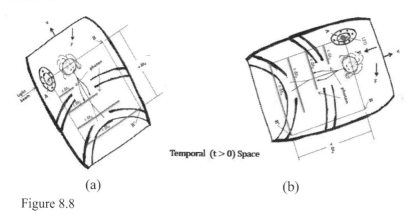

Figure 8.8

In each figure above, a spaceship is traveling close enough to the surface of the earth to be subject to its gravitational pull. Fig. 8.8(a) depicts a spaceship traveling perpendicular to the path of a light beam as it enters the window. Fig.8.8(b) depicts a light beam emitted by an LED, which is attached to the window of a spaceship. Since both spaceships are traveling in different directions with respect to the direction of earth's gravitational pull, gravity should affect each light beam in a different manner. (Note that in both figures, v is the velocity of the spaceship traveling close to speed of light (c), and F is the gravitational force acting on the spaceship.)

Although both spaceships are traveling obliquely to the force of gravity, both light beams remain unaffected by the force of gravity (even if acceleration creates the illusion that they are affected by it). Thus, whether the light beam emerges from outside the spaceship or inside of it, its path remains independent of the force of gravity. Once again, if Einstein had used a temporal (t > 0) subspace when formulating his general theory, he might not have come to the erroneous conclusion that gravitational force can cause light to curve and bend as it travels through space.

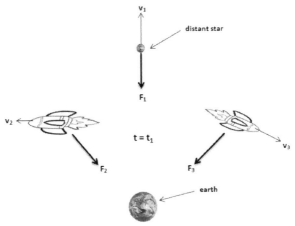

Figure 8.9 Temporal (t > 0) space

In figure 8.9, we see two spaceships orbiting earth while traveling in different directions. Both spaceships are close enough to earth to be subject to its gravitational pull. At time $t = t_1$, a light beam enters the window of the spaceship on the left, while another light beam emerges from an LED inside the spaceship on the right. At the exact same moment, a light beam radiates from a distant star.

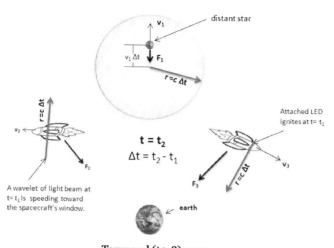

Figure 8.10

In figure 8.10, we can see that a light beam from a distant star travels a radial distance $r = c \Delta t$, where $\Delta t = t_2 - t_1$ representing the time elapsed. At the same time, the light beam entering the spaceship on the left and the light beam emerging from the LED inside the spaceship on the right end up traveling the same distance, $d = c \Delta t$. Neither the light beam entering the spaceship on the left, nor the one emerging from inside of the spaceship on the right is affected by earth's gravitational pull.

Figure 8.11 shows a time-composited diagram of the particle-gravity dynamic whereby the particle exhibits a mass, the photon exhibits no mass, vx is the velocity of the particle traveling in direction x, vF is the average velocity due to gravitational acceleration g, at time $t_2 = \Delta t_4$, and B" is the location of the particle at time $t = t_2$.

Introduction to Physically Realizable Physics

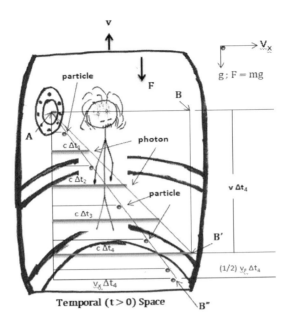

Figure 8.11

Einstein's light bending theory is faulty for the following reason:

It is rooted in Newton's second law, and specifically the equation, $mg = (GmM)/r^2$ where acceleration, g can be written as: $g = GM/r^2$

The theory treats light as if it has mass. Light, however, is a wave and therefore, exhibits no mass.

Assuming that the velocity for an emitted particle is equal to or slightly less than the speed of light, $vx \leq c$. Since every physical particle has a mass m, and mass is necessarily affected by gravity, the path of the particle will be different from that of the light beam.

Thus, gravity bends the path of particle, but not that of either light beam.

Concluding Remarks

When a great scientist comes up with a new theory, we have a tendency to blindly believe in it. Einstein's genius is one of the reasons physicists have become apostles to the religion of time travel and light bending.

There is no physical evidence for time travel and light bending.

Scientific theories should be backed up by physical evidence.

We have forgotten that the pursuit of science is itself, the act of approximating nature's laws, not dictating them. We have forgotten, in other words, that science is never exact.

The light-gravity dynamic is core to Einstein's general theory of relativity. Yet, it does not hold up to physical reality.

Unfortunately, many of today's scientific theories are rooted in Einstein's general theory (for example, Hawking's theories on blackholes and Thorne's theory on wormhole travel). As a result, such theories should be seen as being more in the realm of science fiction than in that of science.

CHAPTER 9

Einstein's Spooky Distance

He knew instant entanglement is spooky.

To see is not to believe?

One of the most misleading principles in quantum mechanics

Since modern physics was developed from Einstein's zero-summed four-dimensional spacetime continuum, we see that Schrödinger's quantum mechanics was also derived virtually from the same time-independent space. For which his wave equation is time independent. This is precisely the reason why his superposition principle is timeless or independent from real-time.

So, what is the superposition principle as I stated, particles or quantum leaps can exist simultaneously and instantly within our time-space. And this is the principle that quantum communication scientists are depending on for the construction of a quantum computer. In other words, without the superposition principle, then there would be "no" quantum computing.

And this is the principle that Einstein had objected most as I will show in this presentation.

Spooky action at a distance

In a letter to Max Born in 1947 Einstein said of the statistical approach to quantum mechanics, which he attributed to Born, "I cannot seriously believe in it because the theory cannot be reconciled with the idea that physics should represent a reality in time and space, free from spooky action at a distance.

The following pages demonstrate how entanglement *does* exist within a limited distance. But the proven entanglement, does not mean that Schrödinger's Superposition principle is correct.

Furthermore, as one has proven that the superposition principle is wrong which is "not" meant that Einstein's relativity theory is correct.

And this "in part", to show that scores of principles and theories are questionable, since foundation of modern physics was established from Einstein's spacetime continuum which is "not" a physically realizable paradigm.

A Classic Entanglement Scenario

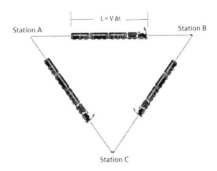

Figure 9.1

For example, within "classical mechanics" regime, we have two trains 1 and 2 of the same length L started the same time from point A. Train1 goes on to point B and instantly turns to point C. Instead train 2 goes on directly toward station C which is a short distance.

From which we see that, if the extra distance from station A to B for train 1 is smaller than the length L of the train, that is $d < L$ we see that train 1 will eventually collide or "entangle" with train 2 at point C. Where L is known as "entangle length" (e.g., correlation or mutual coherence length). From this we see that the entangle distance is limited by the length of the trains L.

However, if the separation between stations A and B is greater than L (i.e., mutual entangle length) of the trains, that is, $d > L$.

Train 1 will not entangle (i.e., collide) with train 2 at point C.

Furthermore, if we assume the length of the trains is infinitesimally short (i.e., $L \to 0$), then we see that it is very "unlikely" these two trains would be entangled, even we assume the distance between A and B is negligibly small (i.e., $d \approx 0$). Of which, these two trains have had entangled at point A, and will entangle all the way to point C and beyond.

Quantum Wavelets Entanglement

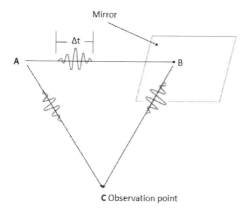

Figure 9.2

However, within "quantum" regime, things change at speed of light. we assume two quantum wavelets started from point A one goes to point B and then reflected toward point C, while the other wavelet started at the same time but goes toward observation point C, which is a shorter distance.

From this, we see that, if extra distance d from point A to B for wavelet 1 to travel is much greater than the pulse width Δt times speed of light c that is, $d > c \Delta t$

Then these two wavelets will "not" be entangled (i.e., meet) at point C, where $c \Delta t$ is the mutual "entangle length" of the wavelet 1 and 2 (e.g., mutual coherence length).

Although we see that quantum entanglement exists within a limited distance. But it does "not" prove that the superposition principle is correct since it is "not" instantaneously entangled. From which we see that, proven Einstein's spooky distance is "false" and is "not" enough sufficient to support the superposition principle as correct. Let me stress that, within quantum regimes particles change at the speed of light, while within classical mechanics domain things change with speed of motion. Yet, instant entangle "does not" exist within our time space, except at current moment (i.e., $t = 0$) of absolute certainty, or within an empty virtual empty space. And this is precisely where all the modern physical principles and theories derived from, unfortunately it is a non-existent subspace within out temporal (t > 0) universe.

In other words, within our time-space perfect entanglement exists at t= 0 absolute moment of certainty of simultaneous quantum leaps. The degree entanglement decreases as distance increases within the entangled length (i.e., c Δt). From which we see that, if the hypothesis is situated within an empty space scenario, instantaneously entangle is possible since empty space has no time and no space. However, as the empty space paradigm is a mathematical timeless space it does not exist within our time space.

From which we see that, the superposition principle does not and "cannot" exist within our universe. Since quantum computing is based on the existent of superposition principle, why we keep promoting it? Is it because Richard Feynman had said so.

Why Modern Physics is so Wrong?

Since the foundation of modern physics was built from a non-physically realizable 4-d spacetime continuum of Einstein, from which practically all fundamental principles, theories and laws were developed, but this paradigm "cannot" exist within our temporal (t > 0) universe.

Secondly, science is supposed to discover but not to create. Since Einstein's spacetime continuum was created, we see that any principle or theory was developed from this paradigm is doomed to be virtual and fictitious or even "fake". Of which, as I see it, we had been burying ourselves within this non-physically realizable paradigm for over a century.

Although Einstein's energy equation was derived from his special theory, it does "not" mean his relativity theories are correct.

Sincere Remarks

Theoretical physics was started from classical mechanics to modern relativistic and quantum mechanics, and it well be into the temporal (t > 0) mechanics. For which it is fair to say that:

Einstein was "not" aware his 4-d spacetime continuum is "not" a physical realizable paradigm, otherwise he would not have developed his relativity theories. Similarly, Schrödinger did "not" know that his superposition is a non-existent principle otherwise he would not have created his fundamental principle.

Given the above, I predict that temporal (t > 0) science will be a new evolutionary mechanics for this 21st century. Yet it remains to be seen [1].

[1} Francis T.S. Yu *"The Nature of Temporal (t > 0) Science: A Physically Realizable Principle"*, 2022, CRC press.

Since modern physics was developed from Einstein's zero-summed four-dimensional spacetime continuum, we see that Schrödinger's quantum mechanics was also derived virtually from the same time-independent space. For which his wave equation is time independent. This is precisely the reason why his superposition principle is timeless or independent from real-time.

So, what is the superposition principle as I stated, particles or quantum leaps can exist simultaneously and instantly within our time-space. And this is the principle that quantum communication scientists are depending on for the construction of a quantum computer. In other words, without the superposition principle, then there would be "no" quantum computing.

And this is the principle that Einstein had objected most as I will show in this presentation.

Spooky action at a distance

In a letter to Max Born in 1947 Einstein said of the statistical approach to quantum mechanics, which he attributed to Born, "I cannot seriously believe in it because the theory cannot be reconciled with the idea that physics should represent a reality in time and space, free from spooky action at a distance.

For which I will show entanglement does exist within a limited distance. But the proven entanglement, does not mean that Schrödinger's Superposition principle is correct.

Furthermore, as one has proven that the superposition principle is wrong which is "not" meant that Einstein's relativity theory is correct.

And this "in part", to show that scores of principles and theories are questionable, since foundation of modern physics was established from Einstein's spacetime continuum which is "not" a physically realizable paradigm.

CHAPTER 10

Can Space Really Curves Spacetime?

It is time or space that moves light.

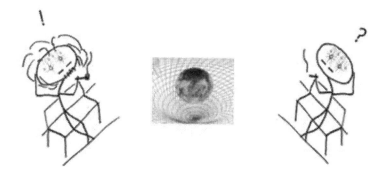

It is space changes with time, but not time changes with space [i.e., space is a function of time, but not time is a function of space since time is a forward variable yet coexists with space]

A few preliminary remarks

Science is not supposed to be deterministic. That is to say, science is meant to discover rather than create the laws of nature.

Science is a way of describing physical reality. As a result, it should correspond to physical reality.

Given the proper evidence and logic, scientific laws, principles, and theories can always be revised, changed and even refuted.

The adage, "you can't get something from nothing," applies to physical reality. In other words, everything in our universe is derived from something else. Everything carries a necessary "price" in the form of time, Δt, and an amount of energy, ΔE. Nothing exists in timeless space, and nothing exists without an expenditure of energy. Nothing, therefore, comes for "free."

Time elapses at a constant and therefore, unchanging rate

Quantum physics falsely posits that time can elapse at varying rates. Practically all the laws and principles of modern science have been derived in some form or another from Einstein's zero-summed 4-D spacetime paradigm.

Einstein's 4-D spacetime paradigm violates both the Second Law of Thermodynamics and the notion that time elapses at a steady and immutable rate. As a result, any hypothesis developed from this paradigm does not correspond to physical reality.

The universe—what I am defining as a physically realizable temporal universe—is a non-zero summed, time-dependent, energy conserving subspace.

Physical reality only exists in the present ($t = 0$) moment. The past ($t < 0$) can be represented by second-order representations such as memories.

(i.e. neural connections in our brain), written records, audiovisual recordings, forensic analyses, computer simulations and stories passed down through oral tradition. Second-order representations, however, do not equal first-order reality.

While memories, stories, written records, audio/visual recordings, geologic formations, forensic evidence and the like can help us to reconstruct and analyze the past (as is done in particular fields such as history, geology, paleontology, archeology, criminal investigation, political economy, political theory and philosophy), they are not equal, from a strictly physics standpoint, to the past. Rather, they are simply partial representations of the past and thus, hints or clues to what actually happened during a particular moment in the past. Second-order representations do not equal first-order reality.

Physical reality, as I am defining it, only exists as a first-order phenomenon. Physical reality therefore only exists in
the present moment.

While the future (t > 0) represents a physically realizable universe, it is nonetheless, uncertain. This is why science should be thought of as a means of approximating rather than determining the future.

Figure 10.1 : Empty space

"Feynman said to Dyson . . . that Einstein's great work had sprung from physical intuition and when Einstein stopped creating, it was because 'he stopped thinking in concrete physical images and became a manipulator of equations.'"
  -James Glick, *Genius, The Life and Science of Richard Feynman**

I have listed below the major differences between the empty space paradigm currently used by both classical and modern physics and the temporal space paradigm that I am proposing:
- The empty space paradigm on the left has been used by the scientific community since at least the time of Newton. The problem is that empty space is unbounded, which means that it entails no space, and consequently, no time and no substance. This means that empty space is a virtual phenomenon which does not exist in the real world. Empty space is, in other words, a fiction, even if it can be derived mathematically.
  Note that like language itself, mathematics is entirely symbolic. As a result, mathematics should be thought of as one of many possible representations of reality rather than reality in and of itself.
- It follows that the more complex and therefore abstract the symbology, the more varied the possible interpretation and application. This means the more complex and abstract the symbology, the more potential there is for misinterpretation, misapplication, and misuse (whether in the case of language or mathematics).
- Thus, even if the internal logic of particular theories (scientific, or otherwise) and forms of mathematics remain consistent, the assumptions on which they are constructed are not necessarily rooted in material reality.

*James Gleick, Genius: *The Life and Science of Richard Feynman*, New York: Pantheon, 1992, p. 244.

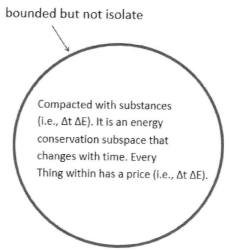

Figure 10.2 Temporal (t > 0) subspace

In contrast to the empty space paradigm, the temporal (t > 0) space paradigm that I am proposing obeys the principle of energy conservation handed down from classical physics. Everything in temporal space consequently carries a "price tag," meaning that matter can only exist within the context of "spending" (i.e. expending) time and energy. The mere existence of any given particle consequently entails the gain or loss of a specific amount of energy ΔE, which in turn, entails an expenditure of time Δt (where $\Delta t \, \Delta E = E_o$, with E_o being total energy in the bounded subspace). Under these circumstances, we can see that for any given particle, the energy contained within it (ΔE) changes naturally with time (Δt). We can see, in other words, that in the temporal space paradigm, the Second Law of Thermodynamics continues to hold true.

Another way to state the above is to say that time entails space and distance entails time. Any particle, no matter how small, consequently has a definite location at any particular moment in time. Everything, in other words, carries a price in the form of $\Delta t \, \Delta E$. In this context, space is always a dependent variable while time is always a perpetually forward-moving independent variable.

Introduction to Physically Realizable Physics

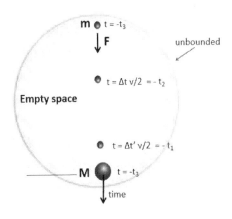

Figure 10.3

Figure 10.3 shows particle m falling toward earth (represented by M). According to standard particle-gravity dynamics, we have the following: $F = mg = G M m(1/r^2)$, where g is acceleration due to Earth's gravity, M and m are the masses of our planet and the particle respectively, and G is the gravitational constant:

- Particle m is accelerating, meaning that with each passing second, it is falls toward mass M at a higher velocity.
- The velocity at which mass m falls can be written as, $v \approx g \Delta t$, where Δt is the section of time that has elapsed.

If we use the standard Empty Space paradigm, we cannot tell which mass is moving. In addition, since space in this paradigm does not depend on the time required to travel across it, the particle can travel any distance instantaneously. There is, consequently, "no price to pay" in the form of time, Δt or energy, ΔE. The assumption here is that under particular conditions, the time required to traverse any given space can be reduced to zero, the space between objects can instantly disappear and the energy needed to bring any two objects together can be reduced to zero.

Despite the fact that truly empty space does not exist, both modern as well as classical physics have been developed from the assumption that it does.

In order to predict the trajectory of a falling particle, there must be a one-to-one correspondence between the passage of time and the location of the particle. For this to be possible, time must exist as a forward-moving independent variable.

Spacetime can only curve if we allow time to become a directionally unconstrained dependent variable.

The General Theory of Relativity rests on the assumption that time is a directionally unconstrained dependent rather than a perpetually forward-moving independent variable.

Just as with most theories and principles, the General Theory posits a timeless empty space platform.

The dynamics of gravity curving spacetime emerges from a space-dependent time.

Notice that time in an empty space paradigm does not reflect the real nature of time, which is to function exclusively as a perpetually forward-moving independent variable rather than directionally unconstrained dependent one.

Time functioning as a dependent variable (as it does in the Empty Space paradigm) violates the necessary independence of time.

Positing time to be directionally unconstrained violates the perpetual forward progression of time.

If we place the same scenario in a temporal ($t > 0$) subspace, we see that the universal clock ticks at the same rate everywhere in our universe. Note that in a temporal subspace, space changes with time. Space, however, cannot change time.

Due to the inevitability of mass annihilation, gravitational pull also changes with time. It is wrong, however, to assume the converse. Time does not change as a result of changes in gravitational pull. There is consequently no way to go backward in time.

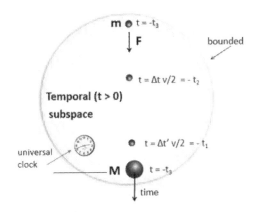

Figure 10.4

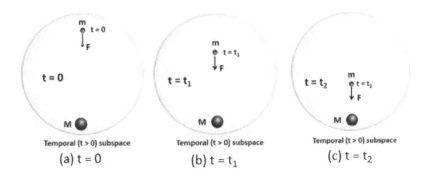

Figure 10.5 We can determine the location of the falling particle. But we cannot go backward in time.

Figure 10.5 shows that the precise location of a falling particle m can be determined by knowing exactly how much time has elapsed [i.e. $\Delta t = t1 - 0$]. (Once again, time is necessarily a forward-moving independent variable, meaning we cannot go back in time. Figure 10.5 shows that the precise location of a falling particle m can be determined by knowing exactly how much time has elapsed [i.e. $\Delta t = t_1 - 0$]. (Once again, time is necessarily a forward-moving independent variable, meaning we cannot go back in time.

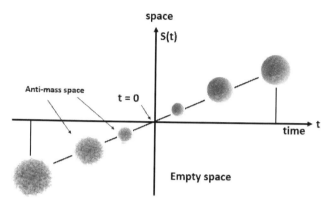

Figure 10.6 Space S(t) as a function of time

Since every physical subspace (or substance) has been created by expending a given amount of energy, ΔE, during the passage of a given amount of time, Δt, each subspace both coexists and changes with time, as represented by ΔE(Δt).

Space, in other words, changes with time. The converse, however, is not true because we cannot alter the passage of time.

Figure 10.7 shows time as a function of spacetime S(t) [i.e., t = S(t)]. According to the equation, space can alter time since the latter has been incorrectly treated as a dependent variable.

In reality, time is not a function of space and cannot act as a dependent variable.

Introduction to Physically Realizable Physics

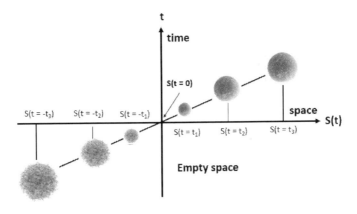

Figure 10.7 Time a function of space S(t)

Empty Space Mathematics

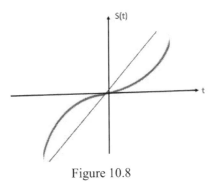

Figure 10.8

There is a one-to-one correspondence between spacetime S(t) and time, t. In other words, spacetime, S(t), is a function of time, t.

Figure 10.9 shows the cone of uncertain space-time which naturally enlarges with the passage of time. The increasing uncertainty is a result of entropy increasing with the passage of time.

Introduction to Physically Realizable Physics

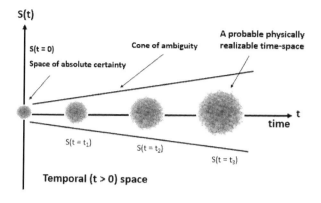

Figure 10.9 Space S(t) as a function of time

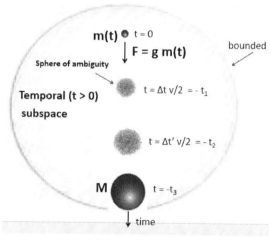

Figure 10.10

In our temporal universe, every subspace, regardless of its size, changes with the passage of time.

In figure 10.10, m is falling toward a stationary M. The further m falls, the larger the sphere of ambiguity. This means that the more time that elapses, the less accurate and precise we can be in determining the location of m.

95

Introduction to Physically Realizable Physics

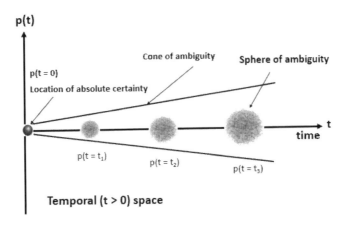

Figure 10.11 Location p(t) as a function of time

As time elapses, there is a proportional increase in uncertainty when attempting to locate subspace m.

There is no one-to-one correspondence between space-time and time. Space cannot, therefore, alter time.

Physical substances change with time as indicated by the Second Law of Thermodynamics.

Conclusion

Einstein's General Theory of Relativity is, in the end, faulty because it is rooted in an empty space 4-D space-time paradigm. One of the main problems with the theory is in assuming that space can alter time. Many misguided theories have emerged from the (false) assumption that space can alter time. Given the fact that time cannot be altered, Hawking's theories on black holes, along with modern physics' concepts of "worm holes" and time travel should all be considered examples of science fiction.

CHAPTER 11

Dark-Age of Our Modern Physics

Theoretical physics has been hijacked by math-oriented physicists.

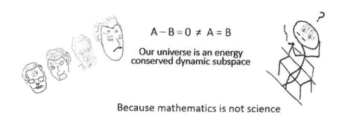

A few remarks

Science is supposed to be physically real; it cannot be virtual or fictitious as mathematics.

Since (fundamental) laws, principles, and theories are to be "discovered" but "not to be created", yet I found the foundation of modern physics was created!

A scientist's dilemma?

Should we continuingly bury ourselves within a created modern science and thrive, or change to a physically realizable science in "bright"?

As a learner, we have lost our "independent" logical thinking for so long and have opted to accept the approval of the others.

Secondly, we "must" transform the fundamental understanding of our modern physics.

Legacy of Mathematical Science

It is fair to say that mathematical science started from Newtonian mechanics, which was developed on an empty space platform, and then moved on to Hamiltonian classical mechanics, to Einstein's relativistic and Schrödinger's quantum mechanics, which are the foundation of physics. However, the background paradigm had not been changed a bit (i.e., empty space) although modern science had changed a lot. And this is precisely the reason why all the laws, principles, and theories are non-physical realizable, as applied directly within our time-space standpoint [1].

Nevertheless, as from an applied and fundament scientist's viewpoint; applied scientists are researching "how" a thing works, but fundamental scientists are searching for "why" a thing works. From which understanding a physical realizable platform is very crucial, otherwise their predictable solutions would be as virtual as mathematics or even fake. And this is precisely the reason why our modern physics has been hijacked by math-oriented physicists, since science needs mathematics.

[1] F.T.Y. Yu, *The Nature of Temporal (t > 0) Science: A Physically Realizable Principle,* CRC Press, 2022.

Where Modern Physics was Started

Foundation of modern physics [2] was started from a four-dimensional spacetime continuum of Einstein (i.e., an empty space paradigm) [3]. It is a zero-summed spacetime paradigm where time is treated as an independent coordinate. Our time space is a non-zero summed energy conservation subspace, where time is coexisted with space as depicted on the right-hand side. From which it is trivial to understand that Einstein's spacetime continuum is "not" a physically realizable paradigm.

[2] A. Einstein, *Relativity, the Special and General Theory,* Crown Publishers, New York, 1961.
[3] F.T.Y. Yu, *Origin of Temporal (t > 0) Universe,* CRC Press, 2020.

Why Modern Science was Created.

Science is supposed to discover but not to create. In view of Einstein's 4-d spacetime paradigm, we see that it is a created time-space. For examples, it is a zero-summed time-space: firstly, it violates the second law of thermodynamics and secondly it is not an energy conservation subspace. It is not a temporal (t > 0) space: it defies the law of entropy and others. From which we see that it is "not" a physically realizable subspace that should be used for any hypothetical analyses.

This is precisely the reason why modern physics is so weird and so wrong [4].

But we had have used this non-physically realizable spacetime of Einstein for over a century!

[4] https://youtu.be/PiBxdhA03pM

Modern science is so wrong, yet why does it work?

For example, young Schrödinger can hit two birds with one stone, but it does not mean that he can hit two birds every time. For which it does not mean that if "one" solution or experiment is positive, all the follow-up solutions are correct. For example, Einstein energy equation ($E = mc^2$) is correct, but it does "not" prove his relativity theories are correct.

Yet, if modern physics is so wrong, why it does work? The answer is that if those applications do not violate the law of our time-space very lightly, then they will work. It is however those theories, principles, and laws derived from Einstein's spacetime paradigm will "not" work. For examples: Einstein's relativity theories, Schrödinger's superposition principle, Dirac's anti-matter, Feynman's quantum electro-dynamics, Hawking's black hole, wormhole time travelling, curving spacetime and many others will not work.

Gravitational law was discovered, but Einstein's theories were created.

As reference to the figure 11.1, can you tell whether the apple is falling, or our planet is moving up. I believe that Newton had told us that the apple is falling. But Einstein has told us it can be "both" ways.

This is exactly the reason why the gravitational law was discovered, yet Einstein's relativity theory was created, since his spacetime continuum was created.

Secondly if you change the embedded empty space paradigm to a temporal ($t > 0$) subspace, where time ticks at the "same" pace everywhere within the universe, you will find out the apple is falling instead of our planet earth moving up. Can you tell me why?

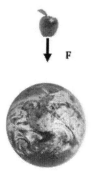

Figure 11.1 Empty space

Observation Limited Resolution

In principle, every observation (i.e., detection) takes a section of time Δt and an amount of energy ΔE to make it happen. Within the classical mechanics regime, an observation is limited by wavelength λ, since time is assumed invariant. For this λ determines the spatial resolution limit. Similarly, temporal resolution within the classical mechanics regime is determined by the velocity v of a particle in motion, for which the temporal resolution is limited by v Δd, where Δd is the size of the object.

However, within relativistic and quantum regimes, things change at speed of light c; every observation is limited by a section of time c Δt, which is the temporal resolution limit, where Δt is the width of a quantum wavelet.

But the foundation of modern physics was developed from Einstein's 4-d spacetime paradigm, from which scores of principles, theories, and laws were derived from this non-existent platform. This is the reason why modern physics is so wrong.

What you see may not be what you should believe

As we accepted that every observation (i.e., detection) takes a section of time Δt to respond. My question is this: can you tell whether the images that you are seeing as depicted in figure 11.2 are simultaneous and instantaneous?

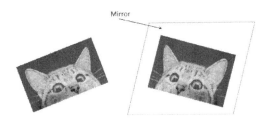

Figure 11.2

A follow-up question is this: assuming you have an image detector that can respond at the speed of light, can you detect these images simultaneously and instantaneously?

If your answer is "no" then how can you legitimize quantum computing, since quantum computing is based on Schrödinger's superposition principle.

Since his fundamental principle was derived with his wave function which is a time independent equation which has "not" existed within our universe. For this, the superposition principle is fake or cannot exist within our time space are simultaneous and instantaneous.

How can we legitimize modern physics?

Since the foundation of modern physics was developed from a non-physically realizable spacetime paradigm of Einstein, from which principles and theories derived from this paradigm are very likely not to be existed within our time-space or even fake.

As a scientist it is my responsibility to let our scientific community know that it is more harmful than beneficial to keep promoting those fancy principles, theories, and laws.

A simple question is that: would you want your children and grandchildren to keep learning those fictious or "fake" sciences?

As a theoretical physicist, if you found that the foundation of modern physics is not a physically realizable foundation, what would you do with what you had learned? I am sure you would turn around to make it more constructive, wouldn't you?

Journey toward special relativity

Recently, a prominent theoretical physicist had hypothesized that a group of spaceships were sent to various corners of our universe, with each spaceship equipped with an atomic clock. As all the spaceships returned to earth, he had anticipated that all clocks should read differently, to prove what he has believed that Einstein's special theory is right.

But my question is this: would you trust the readings from the clocks or the physical reality of what you are seeing. For example, if a clock is registered a million years ago and there is another clock registered as tomorrow, you may wonder why these two clocks are supposed to exist at different times. Yet physically they are coexisted at the "same time" (i.e. at present moment). In other words, would you trust the scientist's hypothesis or the physical realty of what you have observed?

Onus on light bending relativity.

Light waves had been regarded as equivalent to particles in motion by wave-particle dynamics scientists. But a wave is "not" equal to a particle in motion. Since a photon has no mass, gravitational pull does "not" affect light propagation. And this is one of the mistakes that Einstein had made in his general theory. In which he had hypothesized that gravitational force bends light.

Instead, I had found that light bending sensation is due to up lifting of his spaceship hypothesis but not due to gravitational pull. Figure 11.3 shows a time-composited diagram of Einstein's hypothesis. The light beam (i.e. photon) moves across the spaceship at the speed of light and is unaffected by gravitational force. The path of the light beam consequently does not change even though it appears to travel in an arc relative to someone inside the moving spaceship [5].

[5] https://youtu.be/6S8YxnbBe-Q

Introduction to Physically Realizable Physics

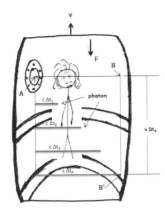

Figure 11.3 Temporal (t > 0) Space

Specious thinking on curving spacetime

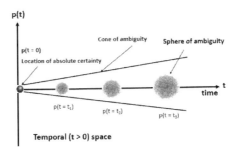

Figure 11.4 Location p (t) as a function of time

Figure 11.4 shows the position ambiguity of a moving particle p(t) within our time space. From which it shows that it does not exist a one – to-one correspondence between particle's position with respect to passages of time. From this we see that that space "cannot" curve spacetime within our universe, since time is a one-way forward-dependent variable with space (i.e. coexists with space).

From this we see that a substance in motion changes with time that follows the law of entropy (i.e. entropy increases naturally with time). In other words, once it had happened within a given time space, we cannot get back that specific time-space. In other words, time is a passaged variable that coexists with it's time-space.

Paradox of a mutual exclusive principle

Empty (i.e. timeless) space and non-emptiness (i.e. temporal) space are mutually exclusive. Then, without the existence of a non-existent (i.e. empty) space, how can we have the existence of our time-space?

In other words, the existence of our time-space needs the existence of timeless space. But timeless space is supposed not to be existed within our time-space. So where has the timeless space existed?

As from our time-space standpoint, timeless (or empty) space is a "virtual" mathematical space. This is precisely why mathematics is "not" physics, yet without mathematics (i.e. timeless) there will be "no" theoretical physics.

CHAPTER 12

Where Dark Matter and Dark Energy comes from?

Our universe is an energy conserved dynamic expanding bounded subspace.

 Entropy is an energy degradation principle; it is reasonable to assumed where dark matter and dark energy comes from

Degraded energy left behind within an energy conserved subspace.

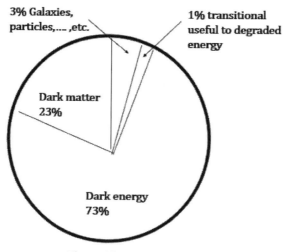

Figure 12.1

Since the universe is an energy conservation dynamic subspace, it is reasonable to assume dark matter and dark energy are the products of degraded energy within our universe.

As a result, I shall start with Maxwell's demon exorcist to the creation of our universe, as from the thermodynamics standpoint (e.g., a law of nature).

Since entropy increases naturally, it is because the boundary of our universe increases with time. Yet, an amount of entropy increased is equivalent to an amount of degraded energy left behind within our expanding universe. For which amount dark matter and dark energy must be the products from the degraded energy that created our universe, since energy within our universe is conserved.

Energy Conservation Subspace

Let me begin with a thermally isolated chamber which is situated within our time-space [i.e., temporal ($t > 0$)] universe] as depicted in Figure 12.2. For this we see that this isolated chamber cannot be empty (i.e., filled with substances) and her total "enclosed energy" is conserved (i.e., an amount of ΔE)

Introduction to Physically Realizable Physics

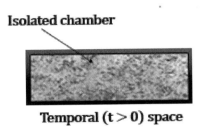

Figure 12.2 shows a thermally isolated chamber.

Since this isolated chamber is assumed thermally equilibrium, of which this chamber cannot do work. For this isolated chamber to do work, if and only if this thermally isolated chamber is perturbed. In other words, is an isolated chamber can be perturbed then it is not isolated or to enlarge the size of the isolated chamber.

Maxwells's demon exorcist

For example, if the isolated chamber is partitioned into two, it can be shown that work can be done by a demon's intervention (i.e., perturbation) as equipped with trapped door shown in Figure 12.3 [1]

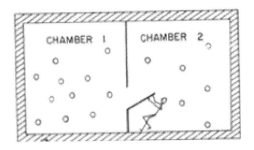

Figure 12.3 shows a thermally isolated chamber operating by demon.

Nonetheless everything within our time-space has a price, for example the demon needs a section of time Δt and an amount of energy ΔE to operate the trapped door. Since as had assumed the demon can see the molecules in motion. Then by sorting higher and lower energy molecules on one side and the other, the isolated chamber can perform work. Nonetheless, the assumed hypothesis must be situated within a temporal (t > 0) space, otherwise there has "no" time for the demon to operate the trapped door.

[1] F.T.S. Yu, *Optics and Information Theory, Wiley Interscience*, New York, 1976.

Every subspace obeys the law of entropy.

Since every subspace must obey the law of entropy, then can a thermally isolated subspace increase its entropy by itself? The answer is yes, if the size of an isolated subspace can be enlarged even without any demon intervention. For example, as an isolated subspace enlarges, its entropy increases with the size increase. And this is precisely why every subspace or substance within our universe, regardless the size, changes naturally with time. And this is the kind of time-space or temporal (t > 0) universe that we are living in. In which we see that, every subspace within our universe is a dynamic subspace that obeys the law of thermodynamics (i.e., law of entropy), otherwise this hypothetical isolated subspace cannot be existed within our universe.

Nevertheless, how can a thermally isolated subspace enlarge her size since it is isolated? It could, for example, if an isolated subspace behaves like a hot air balloon as I shall describe. In other words, it needs an amount of burning fuel (i.e., ΔE) to enlarge an isolated balloon (i.e., subspace).

An energy degrading subspace example

Assuming an isolated subspace within our universe is compacted with hydrogen and oxygen atoms as illustrated in Figure 12.4. By slowly igniting hydrogen atoms with oxygen, that transforms into water molecules.

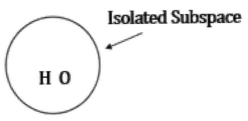

Figure 12.4 show a thermally isolated subspace. It is compacted by hydrogen and oxygen atoms.

But it will take a section of time Δt to produce with an amount of energy ΔE that causes a dynamic expansion of the isolated subspace, as illustrated in Figure 12.5.

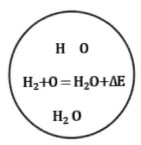

Figure 12.5 shows a dynamic expanding isolated subspace. Within the subspace, it is composed of H, O, atoms, H_2O molecules, an amount H_2 atoms is bonding with O atoms to continuingly releasing amount of energy ΔE(t).

From the above, we see that within the expanding subspace at a given time it contains water molecules (i.e., degraded "chemical" energy) components, with some remaining hydrogen and oxygen atoms (i.e., remaining useful energy), and with a radiating energy [i.e., ΔE(t)]. Where ΔE(t)] is the "work doing" or "fueling" this enlarging subspace [i.e., H_2+O = ΔE(t)]. Where ΔE(t) is an active energy component that pushes outwardly within the subspace. In other words, any physical activity within this isolated subspace is dependent upon this fueling energy ΔE(t) until all the useable H and O atoms have had been bonded or used up. Which is the end-of-life of this expanding subspace as from the chemical energy standpoint.

An energy degrading subspace

Nevertheless, chemical bonding of hydrogen and oxygen atoms will eventually be exhausted those leaves only water molecules behind, in which entropy increasing of the isolated subspace stops. This means that work had completely been done. For this we see that, the chemical life of this burning subspace was started from a lowest entropy state of certainty at time t_1 and ended up at t_3 which is a highest entropy state of certainty. In other words, life of this burning subspace stops as soon its entropy stops to increase.

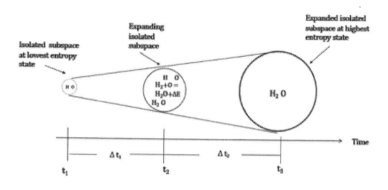

Figure 12.6 shows the dynamic expansion of an isolated subspace. In which depicts of H, O, H$_2$O molecules, and with an associated out-ward energy ΔE due to chemical activity.

Entropy increases naturally with time.

I had shown earlier an example that entropy increases naturally with time, which obeys the laws of nature from a chemical processing standpoint. Similarly, I will show this due to mass-energy equivalent as follows: Based on Einstein's equivalent mass-energy relationship [2], I will give a more accurate equation [3] as given by, $E = \frac{1}{2} mc^2$ where E is the energy, m is the mass and c is the speed of light. From which we see that our universe could have been created from a big bang hypothesis, which is a currently accepted paradigm. But the universe's creation was started within a preexistent temporal (t > 0) space since an empty space model is not a physically realizable paradigm. Of which I shall show the creation of our universe is like an energy conservation isolated subspace as I had described earlier. Nevertheless, our universe was created from the mass to energy conversion of Einstein, which is a divergent energy scenario. From which we had seen that the energy that had created our universe has been continuously degrading since the big bang explosion.

[2] A. Einstein, *Relativity, the Special and General Theory*, Crown Publishers, New York, 1961.

[3] F.T.Y. Yu, *The Nature of Temporal (t > 0) Science: A Physically Realizable Principle,* CRC Press, 2022

A bounded thermodynamic universe

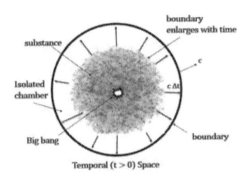

Figure 12.7 shows an expanding isolated subspace within a preexistent temporal space. Where the boundary expands at the speed of light.

Since our universe had been observed to be limited by the speed of light, it is reasonable to assume an isolated subspace is spherical in shape. From which we see that entropy increases as our universe enlarges with time. This is also one of the reasons how our universe was created from a big bang hypothesis, but within a preexisted temporal (t > 0) space as depicted in Figure 12.7. Notice that this paradigm is contrasting with the commonly accepted model that is the big bang was initiated within an empty space [4]. But empty space and time-space are mutually exclusive. In other words, empty subspace cannot exist within our time-space period. But unfortunately, practically all the cosmological analyses were based on an empty space paradigm [5].

[4] M. Bartrusiok and V. A. Rubakov, *Introduction to the Theory of the Early Universe: Hot Big Bang Theory,* World Scientific Publishing, Princeton, NJ, 2011. [5] S. Weinberg, Foundations of Modern Physics, Cambridge University Press; 1st ed., 2021

Entropy is an energy degradation principle.

Since entropy increases with a passage of time Δt, an amount of energy must be degraded within an energy conservation subspace, as given by,

$$\Delta E" = \Delta E - \Delta E'$$

where ΔE is the total "useful" energy, $\Delta E'$ the amount of degraded energy that had left behind. Which it tells that our universe is an energy conservation dynamic subspace where its entropy increases with time from present moment of certainty and beyond, and $\Delta E"$ is the remaining useful energy within our universe. Thus, the larger an isolated subspace increases within a section of time Δt, the smaller the amount of usable energy remains. In other words, more and more degraded energy remains within the dynamic subspace, but "total" energy within the universe (i.e., a dynamic bounded subspace) remains unchanged.

Since the enlarging subspace takes a section of time Δt to increase its entropy, which corresponds to an amount of degraded energy $\Delta E'$ within the section of time Δt that has gone by. From which we see that, everything within our universe has a price tag, a section of time Δt and an amount of energy ΔE to make it happen and it is not free.

An energy conserved dynamic universe.

Figure 12.8 shows that our universe was created by a dynamic expansion of her boundary, which was started about 14 Billion Light Years (BLY) ago by a huge big explosion within a preexisted temporal (t > 0) space.

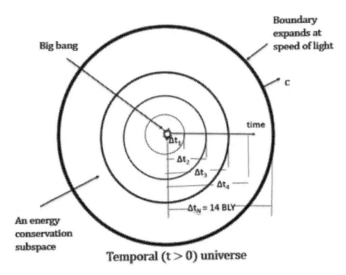

Figure 12.8 shows a time composited expanding universe which shows that energy is conserved at each dynamic state as her boundary expands a speed of light.

Yet it was from a lowest entropy state of a big bang hypothesis. Since everything must change with time, physical reality of our time-space occurs once and only once at present moment t = 0 (i.e., $\Delta t \approx 0$) of certainty. Once an absolute physical reality is emerged, it is immediately gone by with time to the next reality, which cannot be returned. This is precisely the kind of time-space that we are living in. in other words, our universe is an energy conservation bounded dynamic subspace, where entropy continuingly increases naturally with time.

Since our universe is an energy conservation subspace, its total energy is $\Delta E = \frac{1}{2} mc^2$. Of which it is equaled to the amount of current degraded (or decayed) energy plus the amount of current remaining usable energy within our universe as given by,

$$\Delta t_1 \Delta E'(t_1) + E''(t_1) = \Delta t_2 \Delta E'(t_2) + E''(t_2) = \cdots$$
$$= \Delta t_{14BLY} \Delta E'(t_{14BLY}) + E''(t_{14BLY}) = \tfrac{1}{2} mc^2$$

where $E'(t_n) = \Delta t_n \Delta E'(t_n)$ is the amount of degraded energy at $t = t_n$, $E''(t_n)$ is the remaining usable energy respectively at $t = t_n$ the nth dynamic physical reality state, $\Delta E'(t_n)$ is the amount of degraded energy within the section Δt_n at $t = t_n$, M is the mass of the Big Bang explosion, and c is the speed of light, as depicted by a time-composited diagram of Figure 12.8. From which we see that every substance (or subspace) regardless the size, degrades (i.e., decays) with time. Which includes all the elementary particles and non-physical form substances. In other words, everything has a life as from entropy theory standpoint.

A life expectancy of our universe?

As accepted, our universe was created from a big bang paradigm within a preexisted temporal space, and her dynamic subspace expands as its entropy increases with time as shown in Figure 12.9. This means our universe has a life expectancy that is limited by a total conserved energy. Energy cannot be destroyed, but it can be transformed (i.e., degraded or decayed). For this I hypothesize that before our universe has totally exhausted her useful energy, a new big bang creation may have had started somewhere within the huge cosmological time-space. of which we see that, this universe creation by an entropy standpoint is a better physically realizable model since it does not violate any law of nature.

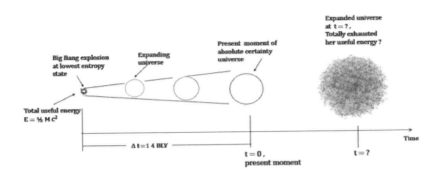

Figure 12.9 shows a time dependent expanding universe. It shows that our universe is an energy conserved dynamic subspace. Her life expectancy is limited by the energy that created our universe.

One and only once physical reality

Physical reality exists once and only once at present moment (i.e., t = 0) of certainty, since our universe is an energy conservation dynamic subspace as shown in Figure 12.9. From which we see that it is impossible for math-oriented scientists to hypothesize that multi universes [6] can simultaneously exist, since physical reality can occur only once. In other words, once a physical reality is emerged, it has no way to get it back. This is precisely the universe that we are living in, yet for year scientists are still fantasizing all those fictious sciences that violated all the laws of nature.

[6] T. Siegfried, *The Number of the Heavens; A History of the Multiverse and the Quest to Understand the Cosmos*, Harvard University Press, pp. 51-61, 2019.

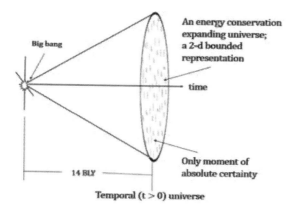

Figure 12.10 shows an expanding isolated subspace within a preexistent temporal space where the boundary expands a speed of light.

Our observable universe

Since every detection takes a section of time Δt and an amount energy ΔE to observe, simultaneity and instant observation depends on the speed of light. Of which the observable universe is limited by a time window Δt of the past to present moment of certainty (i.e., from 14 BLY to current moment $t = 0$). Yet some of the past observable substances (e.g., stars and galaxies) may not exist at the present moment, which gives us a sense of seeing may not believing. Nonetheless, within this observable time-window, we can see the amount of degraded matter and degraded energy that was left behind within our expanding universe as her entropy increases. At the same observation, we are seeing portions of the remaining usable energy and the dynamic degrading energy process within our expanding our time-space. And this is precisely the observable time-window (i.e., Δt) that we have captured with the world's finest telescopes [7]. From which we can observe our cosmological sky from the moment of absolute certainty $t = 0$ to almost at the edge of our universe. This is the physically observable universe that we are seeing.

[7] R. Zimmerman, *The Universe in a Mirror: The Saga of the Hubble Space Telescope*, Princeton Press, Princeton, NJ, 2016.

Back to the energy degradation principle

Since every bit of energy ΔE takes a section of time Δt to degrade, but at the expense of an amount of entropy increased, due to an incremental expansion of our universe, as given by,

$$\Delta E" = \Delta E - (\Delta E_1' + \Delta E_2')$$

where ΔE (i.e., $\frac{1}{2}mc^2$) is the total conserved energy within our universe, $\Delta E_1'$ is the amount of energy that had been degraded from the big bang to the present moment of observation, and $\Delta E_2'$ is the remining useful energy which has not yet been degraded, plus an amount of useful energy in the proceeding state of being degraded (e.g., burning state) which is a dynamic transitional stage from useful to degraded energy form; but it takes a section time to be completed. This is the section of time where work is doing to expand our time-space or the lifeline of our universe, that forces our universe to change with time. In other words, without this transition window, it would not have entropy increased, and we would have no universe and no time. For which the creation of our universe is a continuingly process in transforming usable (e.g., mass) to used (e.g., degraded) energy. In other words, $(\Delta E_1' + \Delta E_2')$ is an anticipated degraded energy at our observation.

Where do dark matter and dark energy come from?

In view of dark matter and dark energy as reported [8], it shows a great in volume within our observable sky as depicted in Figure 12.11 . For this is reasonable to hypothesize that dark matter and dark energy are parts of the energy degradation within our observable sky since our universe is an energy conservation subspace. As we had seen everything within our time-space has a life, it is the energy created in our universe that is conserved. But eventually the energy that created our universe will be degraded with entropy increases. Yet, it is the continuingly degrading part that causes entropy to increase, which gives life to our universe [i.e., temporal (t > 0)] . By which we see that everything within our time-space, regardless the size, has a life.

[8] J. A, Fireman, A. Joshua, M. S. Turner, and D. Huerter, (2008). *"Dark Energy and the Accelerating Universe"*. Annual Review of Astronomy and Astrophysics, 46 (1): 385 432(2008).

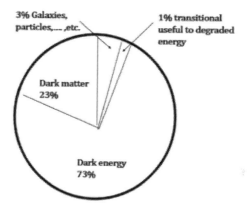

Figure 12.11 shows the distribution of dark matter and dark energy in volume percentage (i.e., %), over our observable sky. Yet, it is the continuingly degrading of useful energy that causes entropy to increase. Which is the life of our universe.

Conclusion

Our temporal (t > 0) universe, unlike the conventional zero-summed spacetime, is an energy conservation dynamic subspace. I have shown that entropy increases with the passage of time, giving rise to a good indication that the dark matter and dark energy are parts of the degraded energy within our universe.

As we have accepted that dark matter and dark energy is equivalent to degraded energy that corresponds to an amount of entropy increase within our universe, we see that doing work is in part corresponding to continuous fueling of the expanding (i.e., lifeline) our universe. Since some observable stars, galaxies, and other substances were no longer existed, the rest of the usable energy are some of the observable substances (i.e., stars, galaxies, gasses, and other usable substances) within our universe.

Once again, the temporal ($t > 0$) universe paradigm is currently one of the closest to the truth model, that obeys the laws of nature. While the commonly used empty spacetime paradigm violates all the laws of nature, for examples the law of entropy, law of time, and others.

Finally let me ask, since we know modern physics is illogical and weird, why have we selected the fake over the real?

CHAPTER 13

Why Einstein's relativistic mechanics is against the laws of nature.

Irony of create relativistic principle.

Principles are approximated yet cannot be created

Einstein's revolutionary theory was developed from a non-physical realizable empty space platform.

So, what is his platform? It is his four-dimensional spacetime paradigm, which is a zero-summed spacetime model that violates all the laws of nature.

What are the laws of nature? To name a few: Law of thermodynamics, law of time, law of entropy and law of energy conservation.

Since everything within our universe has a price tag, a section of time Δt and an amount of associated energy, ΔE is not free.

Einstein was one of the greatest fundamental physicists of all time, but his relativity theory is one of most misleading science theories of the 20th century.

Nevertheless, the fault is due to us, since we knew his relativistic mechanics is so irrational. Yet we opted to follow.

Fundamental principle is to discover but not to create.

Since the fundamental principle of science is supposed to discover from mother nature, it must obey the laws of nature.

What are the laws of nature? Which are the laws that we had discovered that supported the physical reality of our living time-space?

For examples, law of time which is a forwarded dependent variable that coexists with our universe, second law of thermodynamics, law of entropy and law of energy conservation. These are the laws of nature that simply cannot be violated.

In this presentation I shall show that Einstein's relativistic mechanics have violated all the laws of nature. For which it is fair to say that his revolutionary science is as fictitious as mathematics. It is even fake.

Isolated to temporal (t > 0) subspace.

Since the physical realty of a scientific hypothesis relies on a physical realizable paradigm, let me begin with a thermally equilibrium isolated subspace.

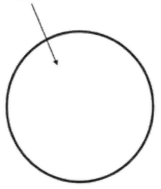

Figure 13.1 Isolated chamber

Since it is a thermal equilibrium isolated subspace, its entropy is at the highest state, but cannot do work. For this subspace to do work, if the size can be enlarged, it takes a section of time Δt to make it happen. This is precisely why isolated subspace must be situated within a temporal (t > 0) space. In other words, without the existence of time it has no way to enlarge the subspace. Secondly, substance and emptiness are mutually exclusive. Therefore, it is wrong to assume that our universe was created within an empty space, as commonly assumed.

Creation of a thermodynamic universe

As we accepted the big bang hypothesis, the explosion must be initiated within a temporal (t > 0) space. Therefore, it follows that our universe is an energy conservation bounded dynamic subspace, which was created within a larger time-space [i.e., temporal (t > 0) space]. Secondly, as the boundary of our universe expands, its entropy increases rapidly from a very low entropy state (i.e., highest degree of certainty) as the boundary enlarges at speed of light. It is important to note that this universe creation is probably one of the closest to the true paradigm, as it obeys all the laws of nature.

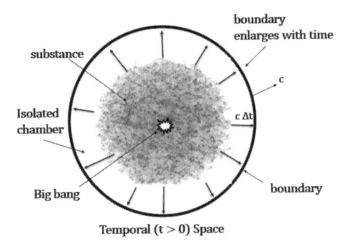

Figure 13.2 Creation of our temporal (t >0) universe

An energy conserved dynamic universe.

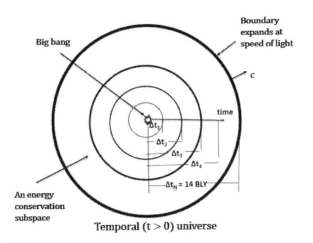

Figure 13.3

Figure 13.3 shows our universe was created by a dynamic expansion. It was started from a low entropy state big bang explosion about 14 Billion Light Years (BLY) ago within a preexisted temporal (t > 0) space.

Since things change with time, physical reality of time-space occurs only once with absolute certainty when t = 0. Once it has gone by, we cannot get it back. This is precisely the universe we are living in. It follows that our universe is an energy conservation dynamic universe, where entropy increases naturally with time.

Since our universe is an energy conservation subspace, its total energy is equal to the amount of degraded (or decayed) energy plus the amount of remaining usable energy as given by the following formula:

$$\Delta t_1 \Delta E'(t_1) + E''(t_1) = \Delta t_2 \Delta E'(t_2) + E''(t_2) = \ldots$$
$$= \Delta t_{14BLY} \Delta E'(t_{14BLY}) + E''(t_{14BLY}) = \tfrac{1}{2} mc^2$$

where $E'(t_n) = \Delta t_n \Delta E'(t_n)$ is the amount of degraded energy at $t = t_n$, $E''(t_n)$ is the remaining usable energy respectively at $t = t_n$. which is the time at nth dynamic physical reality, $\Delta E'(t_n)$ is the magnitude (or amplitude) of degraded energy within a section Δt_n at $t = t_n$, M is mass of the Big Bang.

Therefore, every subspace or substance, regardless of the size degrades (i.e., decays) with time.

One time only physical reality

Physical reality can only exist once at a present moment (i.e., t = 0). It cannot be repeated since our universe is an energy conservation dynamic subspace that changes with time. It follows that it is physically impossible for mathematical scientists to hypothesize that multi universes exist within our time-space.

Because things change with time, physical reality occurs of absolute certainty only at present moment when t = 0 (i.e., Δt = 14 BLY). Once a reality is gone by with time, we cannot get it back. This is precisely the universe that we are living in, that is not against the laws of nature.

Introduction to Physically Realizable Physics

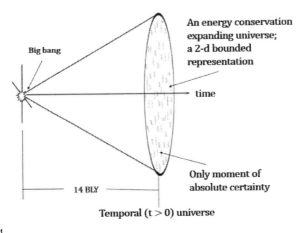

Figure 13.4

Einstein's zero-summed spacetime

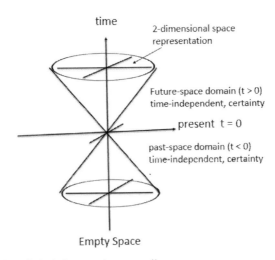

Figure 13.5 : Einstein's 4-d spacetime paradigm

As depicted on figure 13.5 Einstein's zero-summed time independent spacetime continuum. This spacetime paradigm violated all the laws of nature. This is precisely the reason why our modern science is so irrational and so wrong, since the foundation of modern physics was established from this spacetime model.

But the physical reality principle depends on a physical realizable paradigm, for which we see that Einstein's relativistic mechanics, as well as Schrödinger's quantum mechanics, are doomed to be false, since they were developed from this empty space platform.

Fundamental principles are discovered.

One of the greatest discoveries in science must be the gravitational law of Newtonian mechanics, which can be developed from an apple falling from an apple tree, as depicted within an empty space model. Similarly, Einstein could have discovered his special theory if he had assumed the apple is falling at the speed of light.

These two scenarios were situated within an empty space model, but why has the gravitational field been regarded as a discovery, and Einstein's relativity theory is not? It is because Newton had assumed the earth is stationary. But Einstein had assumed the apple and planet earth were moving relatively toward each other. However empty space is a virtual space. It has no time and no space, but we have used this subspace for centuries, not knowing it is not a physically realizable subspace.

Relativistic mechanics fails to exist.

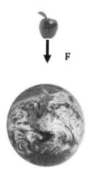

Figure 13.6: Temporal (t > 0) Space Clocks tick at same pace everywhere

Nevertheless, if the same hypothetical scenario is embedded within a temporal (t > 0) space, we see that it is impossible for Einstein to develop his relativistic theories since pace of time moves at the same pace everywhere within our universe.

Since Einstein's relativistic mechanics was developed from his created 4-d spacetime continuum, any principle, theory, or idea developed from his spacetime continuum is doomed to be fictitious and even fake. This includes his special and general relativity theories as well.

Furthermore, if one embeds any of Einstein's hypotheses within our time-space, it is quite easy to figure out that his relativistic hypotheses are not true.

Einstein's spacetime was created.

Since the reality of a fundamental principle is determined by a physically realizable paradigm, it is very difficult for us to continue to support the theory that the foundations of our modern science are physically real. Thus, it is fair to say that any fundamental principle or theory that was developed from Einstein's spacetime paradigm is doomed to be false and even fake.

A reasonable scientist should let our mathematically inclined scientists know that we have been trapped within Einstein's spacetime paradigm for over a century. It is unfair and even harmful to our students, our children and grandchildren to keep learning those fantasy sciences that do not exist within our time space.

Physical realizable principle

Unlike mathematics, science must be physically real. Let me show a simple case of a physical realizable principle. For instance, $\cos(\omega t)$ represents a wave principle. The first question to ask: Is this wave principle a time independent equation? The answer is yes if the principle does indicate it can exist within the negative time domain. However, if your answer is no, then it must be subject to the physical reality of time, which is a forwarded variable that cannot even stop or even change its pace.

Similarly, Schrödinger's time independent wave equation follows the same principle:

$$\psi(t) = \psi_0 \exp[-i\,2\pi v\,(t - t_0)/h]$$

Since it is time independent, it is not a physically realizable wave equation. In physical reality it cannot be used in the negative time domain or even instantaneously within our time space. But why do we keep using it? And this is precisely the reason why quantum mechanics is so wrong.

Are we responsible scientists?

The answer is probably not. Since we were indoctrinated with those fantasy principles, even we knew they were weird, yet we opted to accept the approval of the others. For example, the legendary Richard Feynman once said, "I think I can safely say that nobody understands quantum mechanics". This means that if he does not understand we do not deserve to understand. However, if he would have found out why we do not understand quantum mechanics, then it would have a different story.

Nonetheless, interpretation of a paradox in science means that the paradox does exist. Our time space exists only once in its physical reality. This must be the reason why we were and still are a group of faithful followers, which have opted to accept the approval of the others.

Our modern physics has been polluted by a score of fantasy sciences.

In view of current modern science, there is a score of fantasy ideas (add a bunch of talk show scientists) that has polluted our scientific community. It is fair to say that those fantasy sciences are more harmful than beneficial. In other words, would you want your students, children, and grandchildren to continuing learning those non-existent principles? We have wasted billions, if not trillions of taxpayers' money, which could have been used for other realistic and more constructive sciences.

Furthermore, where did those observable dark matters and energy come from? Our universe is an energy conserved dynamic subspace that changes with time. There is a good reason to relate the observable dark matters and energy with the degraded (i.e., decayed) energy within our bounded universe, since decayed substances remain within our dynamic bounded universe.

Paradox and science are mutually exclusive.

Empty (i.e., timeless) space and non-emptiness (i.e., temporal) space are mutually exclusive. Then, without the existence of non-existent (i.e., empty) space, how can we have the existence of our time-space?

We cannot have both our time-space ($t > 0$) and timeless ($t = 0$) space. Since timeless space is not supposed to exist within our time-space, so where does the timeless space exist?

From our time-space standpoint, timeless (or empty) space is a "virtual" mathematical space. This is precisely why mathematics is "not" physics, but without mathematics (i.e., timeless) there will be "no" theoretical physics.

From this we see that mathematics is not science, but we have let mathematical oriented physicists hijack our science for centuries. Since they had treated mathematics as science, this is precisely the reason why we have so many paradoxes in science.

CHAPTER 14

Myth of Entropy - Boltzmann's Exorcist

Boltzmann's Exorcist without demon intervention

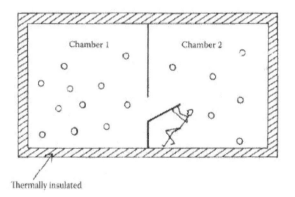

Entropy degrades with time.

Introduction to Physically Realizable Physics

Prelude

One of the most intriguing laws in theoretical science must be the law of entropy. I shall begin with the connect of entropy as from the law of thermodynamics (i.e., one of the laws of nature). Since fundamental principle is supposed to discover but not to create, if a scientific hypothesis is not developed from a physically realizable subspace, it is very likely its fundamental principle is not physically realizable. For example, young Einstein shown his famous energy equation (i.e., $E = mc^2$) is correct, it is by no means that his relativity theory is correct. As I see it, foundation of modern science is in trouble because it was hijacked by mathematical oriented physicists. Since our fundamental physics was developed from an empty space platform (e.g., normally on a piece of paper), and this is precisely the reason that the foundation of modern physics is not real, but virtual as mathematics is. And this exactly why modern physics is so wrong. This is precisely one of my motivations to show that how our time-space (i.e., our universe) was created, as from the principle of entropy. Which is one of the most intriguing laws in classical physics.

Maxwell's Demon

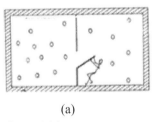
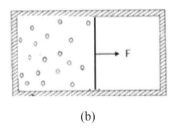

Figure 14.1

Let us hypothesize an isolate subspace filled with particles as depicted in Figure 14.1 (a). Since this isolated subspace is thermally equilibrium, it cannot do work. However, if this subspace is partition into two chambers, equipped with a trapped door operating by Maxwell's demon as shown in Figure 14.1 (a). Since the demon can see each particle in motion, he can sort all the particles into one chamber, as depicted Figure 14.1 (b). By which this isolated is no longer a thermally equilibrium chamber that can do work within chamber. Nevertheless, for the demon to operate the strapped door he needs an amount of energy ΔE and a section of time Δt to make it happen. And this is the reason why the demon unable to provide, since everything within our time-space has a price to pay [i.e., the amount of (ΔE and Δt).

Boltzmann's Exorcist

(a)　　　　　　　　　　　　(b)

Figure 14.2

From Maxwell's demon exorcist one thing we had learned is that, if a thermally equilibrium isolated subspace of Figure 14.2 (a) can be expanded, of which its equilibrium state is broken, like demon intervention of Maxwell. Then within the expanded isolated subspace can perform work as shown in Figure 14.2 (b). From which we see that the expanded subspace includes a huge source of energy [e.g., $E = (½)mc^2$] that vitalizes the dynamic of the subspace. In which we see that continuously enlarging the boundary due increasing its entropy within the thermally isolated subspace, (i.e., Boltzmann's entropy exorcist), means that work is doing the dynamic isolated subspace, yet without the need of a demon intervention.

Entropy: A fundamental law of thermodynamics

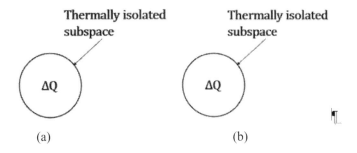

Figure 14.3 (a) shows a thermally isolated subspace is situated within timeless space, (b) shows the same isolated but situated within a temporal (t > 0) space.

Given a thermally isolated subspace as depicted in Figure 14.3, in which an amount of usable energy ΔQ is enclosed. Since it is isolated, its energy within these subspaces is conserved. This is known as the first law of thermodynamics or energy conservation law. Since it is a thermally equilibrium isolated subspace, entropy within this subspace is at its maximum state of certainty. Of which a change from the maximum state is not possible since it is thermally isolated. Nevertheless, if the isolated subspace receives an amount of heat (e.g., ΔQ), an increase of entropy ΔS is given by ΔS = ΔQ/T which is the second law of thermodynamics as from classical mechanics, where T = C + 273 is the absolute temperature, and C is the degree in Celsius. However, this entropy equation is a time independent (or timeless) equation, it cannot be used within our temporal universe. in view of the isolated subspace model of Figure 14.3 (a), firstly it is isolated which has no way to added an amount of heat ΔQ in it. Secondly, it has no time to increase its entropy ΔS, thus it is not a physically realizable model. Nevertheless, as depicted in Figure 14.3(b) it is a physically realizable model where the thermally isolated time-space is situated within a temporal (t > 0) space (i.e., our universe).

Back to classical entropy

In view of the preceding thermally isolated subspace, my first question is that how can one introduces an amount of heat ΔQ into an isolated subspace since it is isolated? For which let me start with the initial amount entropy of absolute certainty within the subspace as given by, $S_0 = Q/T$. Since entropy is also an indicator of complexity but it is also a measure of uncertainty, For example a larger isolated subspace has a higher complexity, as compare within smaller subspace. For instance, its initial entropy of absolute certainty of S_0 is the highest. Although Q/T is a complexity measure, but it is at a certainty state of reality. Nevertheless, as with respect to $S_0 = Q/T$, entropy increase is the lowest. But its initial S_0 is the absolute certainty of physical reality, and this is the equivalent amount of useful energy enclosed. But $\Delta S = \Delta Q/T$ represents an amount of entropy ΔS increased (i.e., equivalent amount of energy degradation) as from $S_0 = Q/T$, ΔS represents an amount of uncertainty increased from S_0. This shows that a portion of original usable energy (i.e., ΔQ) was degraded (i.e., used). This is precisely the reason why entropy is also regarded as an energy degradation principle.

Since Q/T is not a dynamic equation that changes with time, it has no way for the entropy to increase. Nevertheless, if the isolated subspace is allowed to expand by degrading its enclosed useful energy, this subspace must be embedded within a temporal ($t > 0$) space, which allows its entropy to increase with time. This is precisely the reason why energy degrades naturally since isolated subspace enlarges with time.

Since entropy is one of the most intriguing theories in science, yet it is one of the most confusion principles in physics. For example, on one hand is increasing its entropy (i.e., equivalent amount of degraded energy) on the other hand it is decreasing its useful energy.

Entropy is an energy degrading principle.

Since 2nd law of thermodynamics is an irreversible process within an isolated subspace, by which usable energy within the subspace degrades.

This is precisely the reason why its entropy can only increase and cannot be stop. From which we see that every isolated subspace can have only one physical reality that cannot be repeated within our universe. This is why any scientific paradox cannot exist in physical reality. For example, entropy within an isolated subspace starts from an initial amount of absolute certainty (i.e., $S_0 = Q/T$) then increases with time as given by

$$S_0 + \Delta S = (Q + \Delta Q)/T$$

where Q is the total amount of usable energy within the isolated subspace at a specific time of its absolute certainty. But, in principle entropy is continuously to increase with time within any isolated time-space. For which it needs an amount of entropy to increases at the expend of a section of time Δt. and this is the section of time for $\Delta S = \Delta Q/T$ to increase. This amount of ΔS is equivalent to the amount of energy that has been degraded (i.e., ΔQ). For which the amount of usable energy left behind is $Q - \Delta Q$. Thus, we see that total energy within an isolated subspace is conserved, as given by $(Q - \Delta Q) - \Delta Q = Q$, which is the total original energy within the subspace. Since the amount of degraded energy ΔQ is proportional to its amount of entropy increase ΔS, we have shown again entropy is an energy degrading principle. However, energy degrading process is depending on net volume increasing within the expanding isolated subspace, in principle every isolated subspace will die-off at the point of infinity (i.e., $t \to \infty$), as will be discussed later.

Life of an isolated subspace

In view the enclosed useful energy, it degrades continuingly until it exhausted all useful energy enclosed. At that point, the isolated subspace stops to expand, which must be the life of the expanding subspace, since isolated subspace is an energy conserved subspace. For which every isolated subspace well eventually be saturated with degraded energy. That is when the expansion of an isolated subspace stops, which defines the life of an isolated dynamic subspace. Since it is a continuous energy degrading process that supports the life of an expanding isolated subspace, which acts like an internal combustion engine that supports the continuation of life of an isolated space. Since the enclosed subspace has a finite amount of useful energy [i.e., Q = (½) mc²], from which we see that, every substance has a life regardless of its size (e.g., particle). For example, the smaller the mass (i.e., particle) the shorter its life expectancy (i.e., decay), as from entropy energy degradation standpoint.

Nevertheless, energy degradation is a nonlinear monotonical decreasing function. Energy degradation within an isolated subspace will never actually stop, but only at the point of infinity (i.e., $t \to \infty$). In other words, in principle our universe well never actually die, but its dynamic activity will slow down substantially and eventually be subsided at $t \to \infty$!

Classical, statistical, and temporal (t > 0) entropy

Since $\Delta S = \Delta Q/T$ was developed from classical mechanics, which is a time independent (or timeless) deterministic equation. Yet it takes a section of time (i.e., Δt) to make it happen. In other words, an amount of degraded energy ΔQ is needed to increase its entropy (i.e., ΔS). Nevertheless, as from statistical mechanics standpoint, Boltzmann's entropy is given by [1],

$$S = - k \ln p(\sigma)$$

where k is the Boltzmann's constant, ln is a natural log, $p(\sigma)$ is a timeless spatial statistical distribution and σ is a spatial variable. Of which entropy has been transformed from a classical to a statistical mechanics equation.

Since everything within our universe changes with time, Boltzmann's entropy can be modified as given by,

$$S = - k \ln p[(\sigma(t)], t > 0$$

where entropy equation has been presented by a spacetime probabilistic equation {i.e., $p[\sigma(t)]$} and $\sigma(t)$ is an isolated space variable. In which we assert that a larger amount of entropy increases takes a longer section of time Δt to pay off. Nevertheless, entropy in also a measure of uncertainty, for which a longer section of time elapses (i.e., Δt), it is more ambiguous or uncertain to predict.

[1] L. Boltzmann, "*Über die Mechanische Bedeutung des Zweiten Hauptsatzes der Wärmetheorie*". Wiener Berichte. 53: 195–220, (1866).

Boltzmann's entropy exorcist

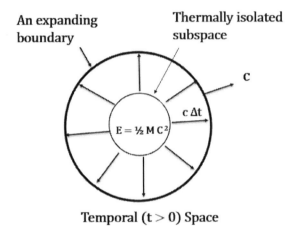

Figure 14.4 Boltzmann's Exorcist without demon intervention.

In view of Maxwell demon exorcist, similarly exorcist without a demon intervention can be achieved by Boltzmann. As I had described in preceding, we had seen entropy can be increased by expanding the isolated subspace without any demon intervention. Of which we had seen, enlarges the size of an isolated subspace is equivalently degrades its useful energy within the subspace. Since isolated subspace is energy conserved, we had shown that entropy theory is also an energy degradation principle.

Now, let us assume a continuingly expanding isolated balloon without any demon intervention, as depicted in Figure 14.4. Since entropy within an isolated subspace will continue to increase as long its size is continually enlarging. This is precisely the reason how our universe is a continuingly expanding subspace of Boltzmann's Exorcist, but without any demon intervention. From which by continuingly degrading its useful energy from a gigantic mass [i.e., $Q = (½)mc^2$], we see that our universe is a self-propelling energy-degrading engine of Boltzmann, without any intervention by a supernatural being.

An energy conserved dynamic universe.

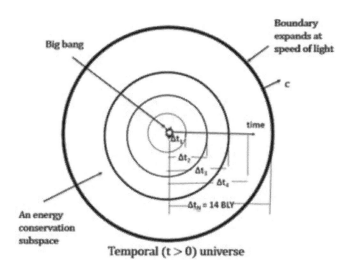

Figure 14.5 A time-composited diagram that our universe is a conserved energy degrading machine as universe is expanding as speed of light.

From preceding we had shown our universe is not a static subspace, but a dynamic energy conservation universe, for which our universe in part is a continuous energy degrading machine as depicted in Figure 14.5. In which a huge energy reservoir as derived from a big bang hypothesis is continuingly degrading that creates our dynamic universe. Since our universe is an energy conservation subspace, its total energy is equal to the amount of degraded energy plus the amount of remaining usable energy as given by;

$$\Delta t_1 \, \Delta E'(t_1) + E''(t_1) = \Delta t_2 \, \Delta E'(t_2) + E''(t_2) = \cdots$$
$$= \Delta t_{14BLY} \, \Delta E'(t_{14BLY}) + E''(t_{14BLY}) = \tfrac{1}{2} mc^2$$

where $E'(t_n) = \Delta t_n \, \Delta E'(t_n)$ is the amount of degraded energy at $t = t_n$, $E''(t_n)$ is the remaining usable energy respectively at $t = t_n$, which is the time at nth physical reality, $\Delta E'(t_n)$ is the magnitude (or amplitude) of the degraded energy within a section Δt_n at $t = t_n$, M is mass of the total energy. From which we see that, every subspace or substance, regardless of the size degrades (i.e., decays) with time.

Life expectancy of our universe

Since thing changes with time, it has a life. If there is no time, then there is no life. But it needs an amount of energy ΔE with a section of time Δt to create our universe [2]. For example, an isolated balloon contains an amount of carbon and twice the amount of oxygen. If we let them slowly burns within the balloon, this balloon will expand continuingly until it exhausted all its usable carbon and oxygen molecules. But the total amount of the remained (i.e., degraded) energy of CO_2 is the same as from usable chemical energy standpoint.

Similarly, our universe has a life as from energy degradation standpoint. Since all its useful energy [i.e., $E = (\tfrac{1}{2})mc^2$] well be totally degraded, we anticipate our universe will eventually seize to expand. Which must be the defined of life expectancy of our universe. At that stage, we anticipate the dynamic activities of our universe seized to exist. But in principle our universe will not be totally death, but only at the point of infinity (i.e., $t \rightarrow \infty$). It is however its dynamic activity will be continuing slowly down, as its rate of energy degradation reduces to zero at the point of infinity. Which leaves behind all its degraded energy within a grater cosmological time-space that our universe once embedded in. yet, before its dynamic activity completely dissipated, I anticipated a new creation might have taken place. And this the life of our universe as from entropy (i.e., thermodynamics) standpoint.

[2] F.T.S. Yu *Origin of Temporal (t > 0) Universe: Correcting with Relativity, Entropy, Communication and Quantum Mechanics, Chapter 1*, CRC Press, New York, 1 -26(2019),New York.

Entropy and information

With reference to Boltzmann's entropy and Shannon's information [3] as given by respectively,

$$S = -k \ln p(\sigma), \sigma(x, y, z), \quad I = -\log_N p(n), n = 2, 3, \ldots N$$

where $p(\sigma)$ is a timeless spatial probabilistic distribution, σ is a timeless space variable, and $N = 2, 3, \ldots$, are the independent numbers of possible consequences. In view of these two equations are basically the same, for which they can be traded. For example, an amount of entropy can be traded to an equivalent amount of information. However, Boltzmann's entropy and Shannon's information were presented in timeless classical forms, strictly speaking they cannot be used within our temporal (t > 0) space. Yet, they can be presented in temporal-spatial forms as given by,

$$S = -k \ln p[\sigma(t)], \quad I = -\log_N p[N(t)], t > 0$$

Since every bit of information or entropy is created by physical substance, their probabilistic representation $p[(\sigma(t)]$ or $p[N(t)]$ changes with time. Nevertheless, an amount of entropy, or information, it equivalent to a cost by means of a section of time Δt associated with an amount of energy ΔE, and they are not free. For example, a book has 106 bits of information, but it takes a section of time and amount of energy to create. But there are many boobs have the same number of bits. Similarly, the same amount of entropy does not cannot guarantee to produce the same physical substance, but it is a minimum cost that must pay.

[3] C.E. Shannon, *The Mathematical Theory of Communication*, University of Illinois Press, Urbana, IL, 1949.

Distinction between entropy and information

Since an amount of entropy is equivalent to an amount of energy degradation (i.e., ΔQ), or can be traded with equivalent amount of information. But neither entropy nor information does not show its spatial content. Without spatial content representation either information or entropy cannot exist within our time-space since we are dealing with physical reality science. For which information or entropy principle must represented with a spacetime variable [i,.e., $\sigma(t)$]. For example, a book has a spatial information content of one million bits, which is equivalent to an amount of entropy or equals to an amount of degraded energy. Since entropy is a timeless quantity, we begin the creation of our universe as started from a huge explosion by an isolated substance (i.e., $E = \frac{1}{2} mc^2$) as depicted in Figure 14.6:

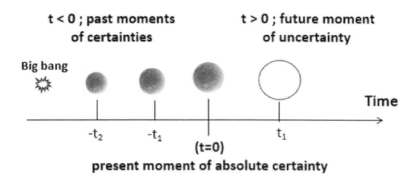

Figure 14.6 shows creation of our universe, where past consequential universes of certainty, present universe of absolute certainty, and future universe of uncertainty.

From the above, we see that our universe was started from a lowest entropy of maximum certainty substance. Since our universe is a continuingly updated physical universe that follows the principle of entropy, every consequential physical appearance of the past cannot be repeated, since time is a forward dependent variable. In which it has once only physical reality.

Our dynamic universe

Capacity of information or amount of entropy is dependent on the size of its isolated subspace. Since boundary of our universe expands at speed of light, our universe is a temporal (t > 0) stochastics subspace that obeys Boltzmann's entropy principle (or laws of thermodynamics). For which every consequential physical reality is by a continuingly updated processing, with its energy degrades with time. In other words, physical realizable subspace (i.e., our universe) changes with a temporal space probability as given by,

$$\int p[\sigma(t)]\, d\sigma(t) = 1,\ t > 0$$

which is a normalized conservative equation, where subspace $\sigma(t)$ enlarges with time.

Since $\sigma(t)$ enlarges at speed of light, we have $\sigma(t=0) = 1$ to $\sigma(t=\infty) = 0$, for which we see that,

$$p[\sigma(t=0)] = 1,\ p[\sigma(t=\infty)] = 0$$

Where the peak of $p[\sigma(t)]$ decreases rapidly from 1 of absolute certainty at t = 0, and then decreases rapidly and then slow down toward to the end of its life [i.e., $p[\sigma(t)] \to 0$] with an absolute uncertainty at the point of infinity $t \to \infty$, as depicted in Figure 14.7. However, its total value of $p[\sigma(t)]$ over $\sigma(t)$ equals to 1, represents that $p[\sigma(t)]$ is a conservative equation.

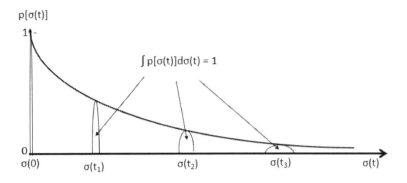

Figure 14.7 depicts a temporal probabilistic distribution of $p[\sigma(t)]$. In which it shows that $p[\sigma(t)]$ is a normalized conservative equation.

A stochastic temporal universe

Since our universe is an energy conservation dynamics subspace, its creation was started from a relatively small subspace with a gigantic energy reservoir at $t = -14BLY$ ago, where its total energy enclosed is given by,

$$E[\sigma(t = -14BLY)] = (½)mc^2$$

As the size of our universe expands to infinite large [i.e., $\sigma(t \to \infty)$ its total enclosed energy is conserved, but its useful energy profile reduces almost to zero, as shown in Figure 14.8 is given by,

$$E[\sigma(t \to \infty)] \to 0$$

But its total enclosed energy [i.e., remaining useful plus degraded energy] is conserved. For which its remining useful energy profile {i.e., $E[\sigma(t)]$} follows Boltzmann's distribution as given by,

$$\int E[\sigma(t)] \, d\sigma(t) = (½)mc^2$$

which is an energy conserved equation, as depicted in Figure 14.8.

Introduction to Physically Realizable Physics

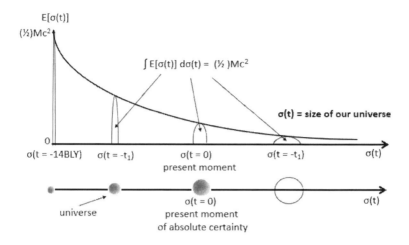

Figure 14.8 shows the dynamic of its energy degradation as our universe expands at speed of light. In which energy profile drops with the size of our universe enlarges.

From this, we see that its useful energy decreases rapidly right after the explosion as its subspace σ(t) enlarges, but slowly its profile decreases to zero when σ(t) approaches to infinitely large. which is approaching the end of her life. At the point, our universe' dynamic activity is very weak, yet the expansion of her boundary remains the same at speed of light. Which is like a ripple vanishes on the surface of a water pond.

An energy degraded universe

In preceding, we had seen absolute certainty happens only once at present moment spacetime (i.e., t = 0) which cannot be repeated. Since entropy increases as the size (e.g., volume) of an isolated subspace enlarges, which is started from the lowest entropy state of absolute certainty. From which we had seen that the rate of entropy increases very rapidly at earlier state of creation as its boundary enlarges at speed of light. The rate entropy increases is due to the ratio of net volume gain (i.e., Δv) to the size of its certainty subspace of V before the expansion (i.e., $\Delta v/V$).

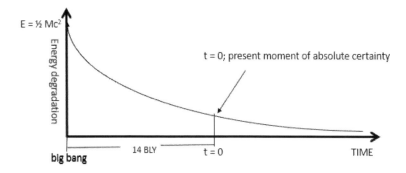

Figure 14.9 shows a hypothetical energy degradation curve within our universe as function of time, instead of size of our universe.

Since the ratio $\Delta v/V$ decreasing rapidly as size of our universe becomes larger and larger, its entropy increases tend to be smaller and smaller as plotted in Figure 14.9 . Where we see energy degrades very rapidly right after the big bang (i.e., $E = \frac{1}{2} Mc^2$) and then decreases monotonically. On which we see rate energy degradation will eventually die down at the point of infinity (i.e., $t \rightarrow$). This is where our universe exhausted all its useful energy to be degraded. Nevertheless, there might have a new universe started to create within the vast cosmological sky, that remains to be hypothesized.

Entropy increases naturally.

Since bigger an isolate subspace higher its information capacity but lower its certainty. Which must be the myths of entropy. In view of the past universes, they were the consequential physical realities of the past as depicted in Figure 14.10 . As our universe enlarges with the speed of light due to energy degradation as hypothesized, we see every consequential universe has only once physical appearance that cannot be repeated, since time is a forwarded variable coexists with space. For which I might have convinced that temporal ($t > 0$) universe is one of the closest to truth model that should be used, otherwise we well forever trapped within the fantasy land of Einstein's universe which has no time and no space [i.e., ΔE(t)] to pay.

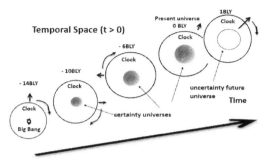

Figure 14.10 shows how our universe changes with time. Which is an irreversible process that follows the myth of entropy.

Conclusion

One of the important tasks for a mathematical oriented physicist is to discover the fundamental laws, but not to create them. For which I found practically all the laws of physics were created by mathematical oriented physicists, where entropy theory is one of them. In this presentation, I had shown our time-space (i.e., universe) can be developed from energy degradation principle of entropy. Since our universe is an energy dynamic subspace, its life expectancy is limited by the amount of energy that created it. From which we see that, smaller mass has a shorter life expectancy. Since there is a similarity between entropy and information, otherwise information theory would be difficult to apply in science. Yet either an amount of entropy or quantity of information does not represent the real physical reality of the substance but as a cost. I had also shown that within our universe it has only one physical reality that occurs only at current moment of absolute certainty. Since our modern physics had been hijacked by mathematical oriented physicists if not for decades it must be over a century. This is about time for us to search for a physical realizable science, otherwise we will forever be trapped within a non-existent fantasy land of modern science forever.

CHAPTER 15

Schematically Disproves Bogus Modern Principles

Einstein's Dice and Schrodinger's half-life Cat

The art of the two greatest physics conspirators

Two universe models, one is away from the truth and the other is closer to the truth. Which one would you select?

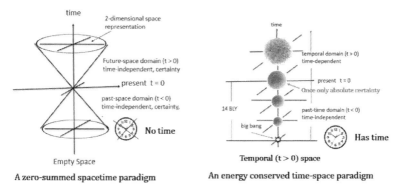

Figure 15.1

Prelude

A picture is worth more than thousands of words since spatial information relies on a light carrier while temporal information depends on sound. By which a schematic presentation is worth more than tens of equations. The fault of modern physics is rooted by its physical realizable platform, but not by mathematics since mathematics is not physics. For which I shall disprove our modern physics schematically instead of mathematically. But without mathematics, analytical physicists have nowhere to go. This is precisely why our analytical physics was stolen by math-oriented physicists.

Nevertheless, every hypothetical principle was started from a schematic representation which had assumed things are situated within an empty space, as we had learned from every elementary science textbook. And this is precisely where the foundation of our physics was developed. Even though the empty space paradigm is a non-physically realizable platform, yet it is from which Newtonian mechanics to relativistic and quantum mechanics were developed. Modern physics is weird, but we had never questioned the legitimacy of those principles because they were developed by great scientists. Yet it was the weirdness of modern physics that I had recently shambled. It is rooted by the empty space platform that we had used for centuries. Since analytical physicists need mathematics, this is precisely the reason why our modern physics was hijacked by mathematics. Nonetheless, the foundation of science is to discover, but mathematics is to create.

Empty space paradigm

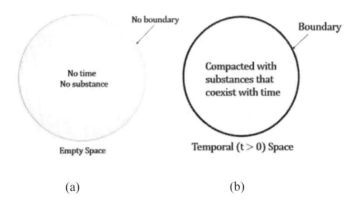

(a) (b)

Fig. 15.2 (a) depicts an empty space paradigm, (b) depicts a temporal space paradigm.

Schematically an empty space is depicted in Figure 15.2 (a), which it has no time, no space, no substance, and no boundary. Yet, math-oriented physicists can implant substance, boundary, and time in it. This is precisely where the root of our modern physics was developed.

Figure 15.2(b) shows a temporal (t > 0) space where space coexists with time. It is an energy conserved dynamic subspace where time advances at the same pace within the expanding universe. Since every subspace changes with the same pace of time, every subspace or substance takes a section of time and an amount of energy [i.e., $\Delta E(\Delta t)$] to create. In which, empty space cannot coexist within our universe, since temporal and timelessness are mutually exclusive (i.e., temporal exclusive principle).

Thus, empty space is not a physically realizable paradigm that should not be used to develop any fundamental principle. Unfortunately, we had used it since the dawn of our analytical science.

Einstein's 4-d space-time is not a physically realizable paradigm.

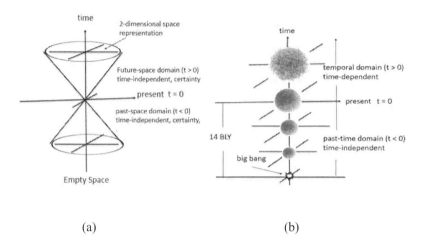

(a) (b)

Figure 15.3 (a) shows a zero-summed space-time continuum, (b) shows a nonzero-summed time-space paradigm

Aside from the temporal issue, a zero-summed space-time paradigm violates the energy conservation law (i.e., 1st law of thermodynamics). For example, Einstein's zero-summed space-time continuum as shown by Figure 15.3(a) was developed from an empty space platform; this is precisely why his relativity theory is so wrong.

A physically realizable non-zero summed temporal space is depicted in Figure 15.3 (b). If any scientific hypothesis is developed with this paradigm, it is very likely that its principle will be physically realizable. Two of the most important pillars of modern physics must be Einstein's. relativity theory and Schrödinger's quantum principle. Yet, both committed the same error, because they were developed on the same empty space platform of Fig, 15.3 (a) which has no time.

Fake relativity theory!

It is from a classic textbook example where Einstein's special theory can be developed as depicted in Figure 15.4. Aside that it is not a physically realizable paradigm, we see that the light beam traveling up to the stationary reflector, and with the one going to the moving reflector are the same, since within an empty space it does not have time.

However, if the same hypothesis is developed within temporal space as shown in Figure 15.5, we see that a tick of the clock will not send the photonic particle up to the upper moving platform. In which we have schematically proven that special theory of relativity is false.

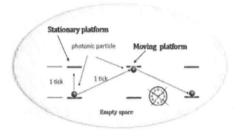

Figure 15.4 show a classical derivation of Einstein's

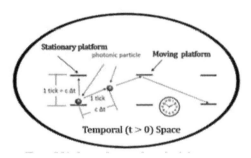

Figure 15.5 shows the same hypothesis but develop within a realizable time-space paradigm.

Counterfeit general relativity theory

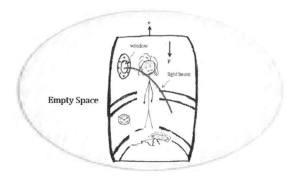

Figure 15.6 depicts Einstein 's bending light postulation as embedded within a timeless space.

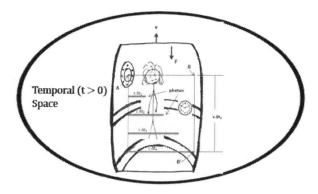

Figure 15.7 shows no bending light due to gravitational field, if the same hypothesis is developed within temporal space paradigm.

Particle-wave dynamics are equivalent but not equal since the gravitational force cannot bend light. This must be one of the major mistakes of his general theory, since he had wrongly assumed that gravity bends light as hypothesized in Figure 15.6.

If the same scenario is situated within our time-space as depicted in Figure 15.7, one can schematically show that the gravitational field does not bend light since electro-magnetic wave propagation is independent from the gravitational force. For which one can schematically show that the bending light sensation is due to the upward lifting of the spacecraft, but not because of gravity, as depicted in Fig.15.7.

Nonsense superposition principle

Since instantaneity is time and simultaneity is space, for which Schrödinger's superposition principle is a time coexisted with space principle. Besides, instantaneity is a physically unachievable quantity since it has no section of time to pay (i.e., $\Delta t = 0$). By which his principle had violated his own theory since he treated time as an independent variable from space. Nevertheless, the superposition principle is the core of quantum mechanics. In other words, without the superposition principle there would be no quantum mechanics. As referred to in Figure 15.8 (a), two wavelets were simultaneously emitted from point A; one goes to point B and then reflected toward point C. But the other wavelet goes toward directly to point C. Since the hypothetical scenario is situated within an empty space, two wavelets will certainly meet at point C simultaneously and instantly. However, if the same hypothesis is situated within our time-space which has time as depicted in Figure 15.8(b), then these two wavelets will not meet at point C instantly, and very unlikely simultaneously since $d = c \Delta t$ is very small, where c is the speed of light.

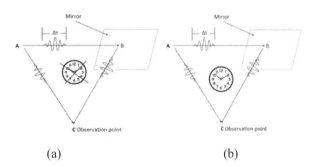

(a)   (b)

Figure 15.8 (a) depicts a scenario of two simultaneous quantum wavelets traveling at different paths within an empty space, (b) shows the same scenario, but it is situated within our time-space.

Timeless half-life Cat

Young Schrödinger can hit two birds with one stone, but this is by no means that he can hit two birds instantly and simultaneously every time. For which a great scientist created a half-life cat, it has intrigued his quantum world for decades. When we pick a half-boiled egg, we cannot tell if it is hard boiled or soft soiled until we crack it open. But if we had forgotten to open it for over a month, would you anticipate the egg would be hard boiled? Since Schrödinger's cat was hypothesized within an empty space paradigm as depicted in Figure 15.9(a), nobody can tell if the cat is either death or alive. Firstly, his box is opaque and secondly his hypothesis is situated within an empty space. These are the reasons that we cannot tell if his cat is either alive or death before one opens his box. On the other hand, consider his box hypothesized within a temporal space as shown in Figure 15.10. Even though it is an opaque box, the cat's life had been determined before one opens the box. From which, it is not because of one's observation that collapses his great principle. But why do we keep promoting this fake principle?

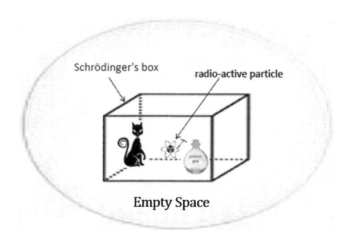

Figure 15.9 depicts Schrödinger's box situated within an empty space scenario.

Introduction to Physically Realizable Physics

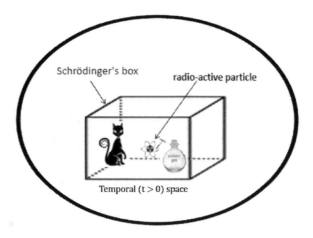

Figure 15.10 shows the same Schrödinger's box but situated within a time-space scenario.

A nonexistent wave equation

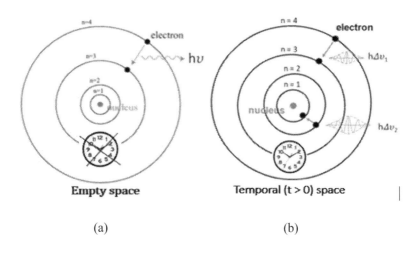

(a)   (b)

Figure 15.11 (a) shows Bohr's atomic model situated within an empty space paradigm, (b) depicts a multi-quantum leap Bohr's atomic model but situated within our time-space.

One of the most important equations of Schrödinger's quantum mechanics must be his wave equation as given by,

$$\psi(t) = \psi_0 \exp[-i\, 2\pi\, v\, (t - t_0)/h]$$

which is the legacy of Hamiltonian classical mechanics but adapted to Bohr's quantum leap as depicted in Figure 15.11(a). Since Bohr's atomic model is situated within an empty space, and as it is not a physical realizable model, his quantum leap is not limited by our time-space. This is precisely the reason why Schrödinger's wave equation is not a physically realizable equation. However, if Schrödinger's hypothesis had adopted Bohr's model situated within a temporal (t > 0) space shown in Figure15.11(b), then his wave equation should have had limited by time (i.e., exists only after excitation) as given by,

$$\psi(t) = \psi_0 \exp[-i\, 2\pi\, v\, (t-t_0)/h],\, t > 0$$

As it is a time dependent equation, this wave equation must be an energy conserved (i.e., time and band limited) equation to legitimize it as a physical realizable wave equation.

Principles cannot be created.

Fundamental principles are supposed to be discovered, not created. With reference to Figure 15.2 (a), a scenario of an apple above the ground is hypothesized within an empty space. Since Newton had assumed our planet is stationary, this must be the reason why he had discovered his gravitational law. With the same empty space scenario, Einstein saw it differently. Of which he hypothesized that it could be either the apple falling, or our planet moving up. Yet Schrödinger disagreed with both postulations; he had concluded that the apple is neither falling nor the planet is moving up, but not until it has happened. These must be the reasons why Newton discovered his gravitational law, Einstein created his relativity theory, and Schrödinger developed his superposition principle. Nevertheless, consider the same hypothesis situated within our time-space as depicted in Figure 15.2 (b). Since within our universe has time (i.e., time-space), we see that it is a serious mistake either to hypothesize that our planet is moving up or that it must wait until we see the apple falling.

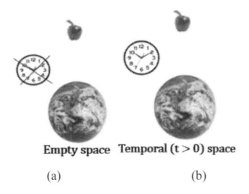

Empty space   Temporal (t > 0) space
(a)   (b)

Figure 15.12 (a) shows a timeless space postulation, (b) depicts the same scenario but situated within a time-space.

Bogus empty space universe

The greatest bogus hypothesis of our universe's creation must be that the big bang creation was started within an empty space, as depicted in Figure 15.3(a). Since empty space is a virtual space, it violates all the laws of nature. Any principle developed from this emptiness cosmological space is doomed to be fictitious and fake.

Nevertheless, our universe is an energy conserved dynamic subspace, as depicted by a time-composited diagram in Figure 15.3(b). It shows that our universe is a continuingly energy degrading process. in other words, it is a continuingly degrading energy process (i.e., from $E = \frac{1}{2} mc^2$) that creates our universe. For which the life of our universe has begun as energy degradation started. But this energy degradation process (i.e., entropy increasing) is continuingly without stopping until it has exhausted all its useful energy that created our universe. In which all the degraded energy will eventually be left behind within the vast cosmological space where our universe once was embedded. From this, our universe has a lifetime. By which every substance, no matter how small, has a lifetime. In other words, the smaller the mass, the shorter its life.

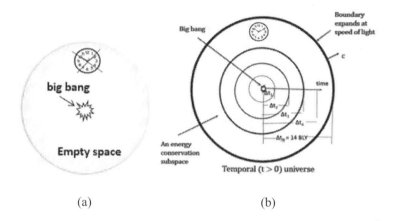

(a)   (b)

Figure 15.13 (a) depicts a well-accepted empty space creation model, (b) shows a time-composited diagram of our universe in which it is the continues energy degrading process that defines the life of our universe.

To see is not to believe

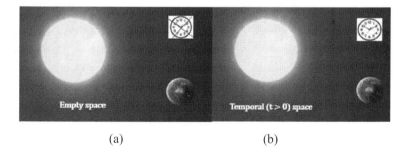

(a)   (b)

Figure 15.14 (a) shows an empty space paradigm, (b) shows a temporal (t > 0) space Model.

The sun rise and sun set tells us that our sun is revolving around us. But Galileo Galilei told us it is the other way around. In view of a commonly used paradigm as depicted by Figure 15.14 (a), Einstein had told us it can be both ways, but Schrödinger told it cannot be both ways until we figure it out. So, which one is right?

However, if the same scenario is situated within our time-space which has time, then we may have to agree with Schrödinger. But we cannot figure it out instantly (i.e., $\Delta t = 0$), since it takes a section of time $\Delta t > 0$ to observe.

Art of the two greatest conspirators

Two of greatest modern physicists of all time had prophesized as I quote: one believed that God does not play dice, and the other believed his hypostasized cat is either alive or death only after his superposition principle collapses, as I cartooned in Figure 15.15. In other words, Einstein's space-time is deterministic, and Schrödinger's binary principle is totally chaotic only after the consequence. Nevertheless, if they had prophesized god's dice and the fate of his cat within a temporal (t > 0) space as depicted in Figure 15.16, they could have changed their prophecies.

Nevertheless, after I had discovered the origin of our modern physics, I cannot reconcile and totally disbelieve that both great scientists that I used to adore were also a pair of the worldwide greatest conspirators of all time.

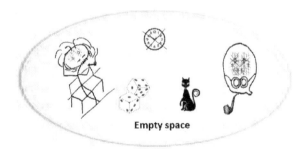

Figure 15.5 shows a pair of greatest physicists prophesizing their thoughts within a timeless environment.

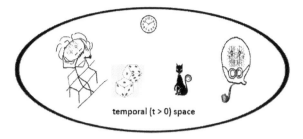

Figure 15.6 shows a pair of greatest conspirators promoting their thoughts within a time-space environment.

Legacy

Modern physics relies on physical realizable principles. Otherwise, it is hard to legitimize since its foundation was established on a timeless fantasy land platform. For which I had discovered that modern physics is irrational, weird, and even fake. But the problem is not the mathematics. Instead, it is rooted on its foundation which was built on a zero-summed empty space platform. Practically all the fundamental principles were developed by math-oriented physicists, and this is exactly why our theoretical physics had been hijacked by math-oriented physicists if not for decades, but it might have been over a century. Since it is more difficult to rebut those principles from a mathematics standpoint, it is easier to disprove them schematically as what I had done in part.

Since science must be physically real which cannot be half real, yet its legacy is dependent on us. Would you like to continue those fantasy principles and thrive or change to physical realizable in bright? Otherwise, we will continuingly be trapped within the fantasy lands of timeless sciences. For example, would you want your students, children, and grandchildren to continuingly learning those nonexistent but fantasy principles of modern science?

CHAPTER 16

From classical to physical realizable Hamiltonian Mechanics

Without being physically realizable, physics becomes filled with fictitious fantasy principles. My question is that when will we return to looking at material reality?

To this day, math-oriented physicists still do not understand why our universe is an energy conservation dynamic subspace!

From classical to physical realizable Hamiltonian Mechanics

Since our universe is not a zero-summed energy subspace, the total energy of a hypothetical machine (e.g., Hamiltonian mechanics) must be conserved (i.e., the 1st law of thermodynamics). For which firstly I will show that classical Hamiltonian mechanics is not an energy conserved machine.

Then, I will formulate a new energy conserved Hamiltonian machine which is physically realizable. Since Schrödinger's quantum mechanics is the legacy of Hamiltonian, but designed for particles that travels at speed of light, his quantum machine is a fantasy timeless machine. I shall reformulate it to becoming a temporal ($t > 0$) quantum machine which will survive within our time-space.

Since Schrödinger's mechanics missed led us for decades, it is about time for us to get back to physical realizable issue, otherwise we will forever be trapped within a fantasy timeless land of quantum world.

Yet, if there was a creation hypothesis, there must be a de-creation hypothesis.

Why mathematics is not science?

Every fundamental equation represents an interpretation, but an equation is not science. For example, a rest particle of mass m is situated within an empty space, as depicted in Figure 16.1 (a). Its total energy due to gravitational pull is its potential energy PE, PE = m g h. Where m it the mass, g is a gravitational constant, and h is the height. which is a timeless equation. Since the scenario is hypothesized within a timeless platform, Hamiltonian classical mechanics is a timeless machine.

But, if the same machine is situated within our time-space as depicted in Figure 16.1 (b), then the particle's PE must be the total Hamiltonian energy. In which we see that as soon the particle falls on the ground, its PE has totally converted into its kinetics energy KE as given by, KE = (½) mv^2. where v is the particle's velocity. For which two possible equations can describe the particle's energy transition from PE to KE as given by, PE - KE = 0, or PE = KE, but one is a zero-summed and the other is a non-zero summed equation. For instance, the zero-summed energy equation suggests that negative energy can exist. While PE = KE, it implies that as particle's PE decreases, there is an equivalently amount of particle's KE increase or vice versa. From which we see that it is not how rigorous mathematics is, but it is the platform that guarantees the physical realizability of its solution.

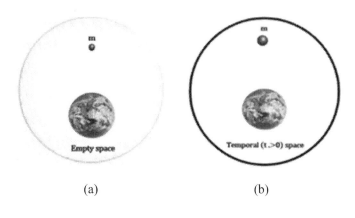

(a) (b)

16.1

Non-realizable classical Hamiltonian mechanics

Figure 16.2 shows how classical Hamiltonian mechanics was developed within an empty space non realizable platform.

Let me refer to a classical Hamiltonian mechanics shown in Figure 16.2, for which Hamiltonian operator takes place, H = PE + p, where p = m v is particle's momentum, v is particle's velocity, and a is its acceleration. From which particle's momentum is independent from its PE. Aside it is not a physically realizable paradigm It's PE has nothing to do with its momentum. Furthermore, if an eigne state wave function ψ is incorporated, the Hamiltonian wave equation takes the form, H ψ: [PE + p] ψ = 0. But it is a not an energy conserved equation, for which ψ is not temporal (t > 0). Let me further note that, justification of introducing an eigne state of ψ is assuming its platform subspace is non-empty. Like electromagnetic wave propagation (e.g., Maxwell equations) was formulated within a non-empty space platform, yet timeless. But it is the time of the platform that moves the wave, but not the wave that changes the time of the platform. In other words, without the existent of time of the platform, ψ cannot propagate with time. Aside the classical Hamiltonian is not a physically realizable operator, but it has a critical issue for the inclusion of particle's momentum (i.e., f = ma) in it since it has nothing to do with the particle's PE. Nevertheless, it was the empty subspace paradigm that made classical Hamiltonian mechanics a timeless fancy machine.

In view of Hamiltonian classical mechanics, where H operator represents a one-dimensional x operator as given by

$$[1], H = - [1/2m] h^2 \nabla^2 + (PE)$$

Firstly, it is not an energy conservation operator, where

$$\nabla^2 = (\partial^2) / (\partial x_i \, \partial x_j)$$

is Laplacian operator.

Secondly introducing an eigenstate wave ψ function is based on particle-wave dynamic idea. Of which Hamiltonian wave equation takes the form

$$[V - (1/2m) h^2 \nabla^2]\psi = E\psi$$

where V is particle's PE, and E is the kinetic energy associated with momentum p = mv and v is particle velocity of p. Nevertheless, wave function of ψ did not tell us whether electro-magnetic or gravitational wave in nature is. Furthermore, from reality standpoint Hamiltonian had hypothesized the scenario within a non-empty yet without time space. In other words, the introduction of ψ is strictly from particle-wave dynamics idea, but Hamiltonian mechanics was developed within an empty space paradigm. Nonetheless, by rearranging preceding equation, we have the following well-known Schrödinger's zero-summed wave equation as given by,

$$\nabla^2 \psi + [(2m)/h^2] [(KE) - [PE]]\psi = 0$$

which is not energy conserved, this is why solution of ψ is independent of time.

[1] D. F. Lawden, *The Mathematical Principles of Quantum mechanics,* Methuen & Co Ltd., London (1967).

A note on Hamiltonian's legacy

Since Schrödinger's machines is the legacy of Hamiltonian classical mechanics, it is fair to say that if classical Hamiltonian mechanics fails to exist within our time-space then it is difficult to legitimize that Schrödinger's quantum machine is physically real. Nevertheless, the major application of Schrödinger's mechanics is restricted to very small particle that travel at speed of light where Bohr's quantum leap was adapted. But Bohr's atomic model was also hypothesized within an empty space platform of which his quantum leap is not time and band limited [2].

In view of the current limit of our physics; it does not permit object that travels at or beyond the speed of light, for which I will discuss application is limited by a package of wavelet (i.e., quantum leap). As for the implementation of Hamiltonian to larger objects, such as to planetary system, in cosmological scale, or others are remained to be seen.

[2] F.T.S. Yu, *"The Nature of Temporal (t > 0) Science: A physically Realizable Principle"* CRC Press, Boca Raton, Fl (2022).

An energy conserved Hamiltonian

Since our universe is an energy conservation dynamic subspace, every hypothesized mechanics within our time-space must obey the law of energy conservation. When Hamiltonian machine is submerged within our time-space as depicted in Figure 16.3, Hamiltonian operator takes the form, H(t): PE(t) = KE(t), t > 0, which is a non-zero summed energy conservation operator which changes with time. From this we see that when particle's PE decreases there is an equivalent amount of the particle's KE increase as given by, H(t): - d[PE(t)]dt = d[KE(t)]dt , t > 0, it shows that total energy of H(t): operator is conserved, which is equaled to particle's original PE at the initial state (i.e., t = 0), PE(t = 0) = m g h(t=0)

Thus, without the existence of time of the subspace that Hamiltonian is embedded in, there would be no transferring of PE to KE with time. This is reason that a physically realizable Hamiltonian must be
an energy conserved operator which changes with time.

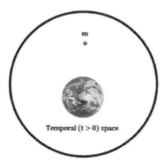

Figure 2.3 shows a Hamiltonian machine is situated within our time space where time coexist with the space [i.e., temporal (t > 0)].

A footnote

Math-oriented Hamiltonian scientists have done one thing right by introducing an eigenstate wavefunction ψ in the analysis. Although Hamiltonian was developed within an empty space model, but the implementation of ψ was surprising cleaver though it must be due to particle wave dynamics standpoint, although Hamiltonian was developed on a no space platform. Yet, particle-wave dynamics indicates that it has empty space medium but timeless. Like many of the classical sciences, emptiness is supposed to have no substance in it and no time, but ψ is a wave dynamics that needs non-empty space to support. The same-like Maxwell equations were developed on an empty space platform, but it shows that EM wave propagates within a dielectric medium but without any constraint within our time-space. This must be the reason theoretical physicists analyze how light propagates from point A to point B, but a fundamental physicist needs to know a section of time Δt with an amount of energy ΔE to move a wavelet from point A to point B. Since within the regime of quantum mechanics, things travel at the speed of light, but it is the section of time with that amount of energy [i.e., $\Delta E(\Delta t)$] that makes it happen, which cannot be simply ignored. Yet we have ignored this price [i.e., $\Delta E(\Delta t)$] if not for decades it must be over a century.

Einstein's space-time continuum could have been avoided.

One of the geometrical coordinate systems that does not have a non-zero-summed feature is spherical coordinate geometry, as depicted in Figure 16.4(a). For which if Einstein's space-time continuum was developed from a spherical coordinate system, he might not have a zero summed spacetime continuum. Of which Dirac's parity physics would not have developed as can be seen in Figure 16.4(b) since every passed realized space-time universe exists only once. Although spherical geometry spacetime is a non-zero summed spacetime, yet it is still empty space issue against its existence within our temporal ($t > 0$) universe. From which we see that it is not how rigorous the mathematics is but it is the physically realizable platform that guarantees a legitimate solution. Nevertheless, the foundation of our modern physics was built on Einstein's 4-d spacetime continuum; no wonder our modern science is so wrong.

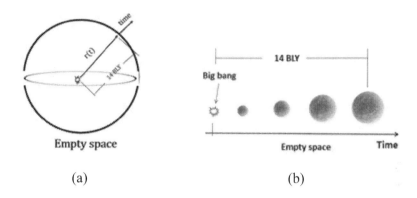

(a)   (b)

Figure 16.4 (a) shows a spherical coordinate diagram of particle's convergent PE within our time-space, (b) shows a spherical coordinate diagram of particle's KE divergent explosion within our time space.

Temporal Hamiltonian: an energy conserved operator

In view of the energy conserved Hamiltonian operator as given by, H: PE = KE which is a timeless equation unless H operator changes with time [i.e., temporal (t > 0)]. From which it tells us H machine works if only if, it is submerged within our time space. In other words, anything exists within our universe coexists with time. For example, when time moves on from present moment (i.e., t = 0) there is an amount of the particle's KE energy increase which is equivalently equaled to an amount of the particle's PE decrease as given by,

$$H(t): - m\, g\, d[h(t)]/dt = (½)\, m\, d[v(t)]^2/dt\, ,\, t > 0$$

As we can see, PE decreases as h(t) decreases, but at the same time particle's KE increases rapidly since particle's v(t) increases. In view of H(t): is a non-zero summed operator, it is a physically realizable machine that obeys all the laws of nature (e.g., law of time, law of energy conservation and others).

A radii Hamiltonian operator

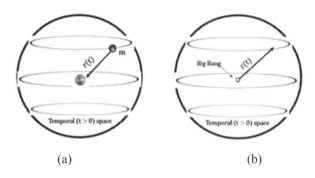

(a)   (b)

Figure 16.5 (a) shows a spherical coordinate diagram of particle's convergent PE within our time-space, (b) shows a spherical coordinate diagram of particle's KE divergent explosion within our time space.

Since Hamiltonian mechanics is a radial machine (i.e., radii dimension) where h(t) can be replaced by of r(t) and particle velocity is an in-warded radii $v_r(t)$ as depicted in Figure 16.5 (a). Of which a radii Hamiltonian takes the form,

$$H_r(t): m\ g\ r(t) = (½)\ m\ [v_r(t)]^2,\ t \geq 0$$

Since particle's PE is a convergent energy vector as given by,

$$PE: -m\ g\ d[r(t)]/dt = -\nabla \cdot S,\ t \geq 0$$

where $[-\nabla \cdot]$ is a radii convergent operator and S is an inward radii energy vector. From which PE represents a gravitational convergent energy vector.

Strictly speaking, the application of Hamiltonian to modern physics is a complicated issue. For examples, as applied to large scale planetary system or to small particles at speed of light. Strictly speaking every temporal (t > 0) mass has a life [i.e., m(t)], the constrains is m(t) degrades with time. Nevertheless, an extreme case due to mass annihilation is given by

$$KE\left(\frac{1}{2}\right)c^2 d[m(t)]/dt = \nabla \cdot S, t \geq 0$$

which shows that particle's KE diverges at speed of light as depicted in Figure 16.5. Where [∇ ·] is a divergent radii operator and S is a radii-energy vector. From this a mass annihilation divergent energy KE vector can be used to simulate the creation of our dynamic universe, as will be seen in the following slides.

Origin of our universes

Nonetheless, the upper limit of particle's KE is reached as its velocity reaches to speed of light. For which the mass must be annihilated into EM radiation, which is equal to the total energy of mass [i.e., E = (½) mc²] as given by,

$$KE: (½) c^2 d[m(t)/dt] = \nabla \cdot S, t \geq 0$$

This is essentially the same equation for the creation of our temporal (t > 0) universe. From which as our universe is expanding at speed of light, it creates a divergent distribution of all substances (e.g., all degraded energy and others) from the center of big bang explosion. For this, it creates a convergent PE due to gravitation force with the expanding universe as dramatized in Figure 16.6. In which we see that more and more degraded substances are produced (e.g., dark matters and others) as the entropy of our universe increases. Yet this creation process will not stop until our universe exhausted all its KE that created our universe. And this at the point of infinity (i.e., at t → ∞), where our universe uses up all its useful energy provided by its KE. But before it reaches the end of its divergent creation, there will be a sufficient convergent PE taking place to reverse the creation process as from all the degraded matters.

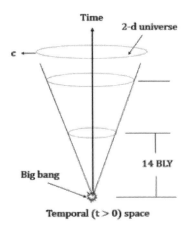

Figure 16.6 shows a divergent creating process primarily dye to mass annihilation. Where its boundary is expanding at speed of light.

De-creation of our universes

Nevertheless, before reaches the end of creating our universe [i.e., exhausted all its useful energy] it will eventually build up a strong convergent gravitational PE from all the degraded matters (i.e., dark matters and others) to overcome the divergent creating process. In which we see that the de-creating process might have started soon after big bang exploded as depicted in Figure 16.7 . From which the dynamic activity within our universe will be subsiding (e.g., entropy increases slowing down) at the end of our universe' life. This means that the remaining useful KE within our universe is at a lowest end. At this point, its convergent gravitational PE will overtake its divergent KE due to an inward gravitational field created by all the degraded matters. Since there are two major energies responsible for creation and de-creation of our universe; divergent KE is the one contributing to creation, but convergent PE must be the one contributing to the de-creation. Since convergent PE has started to build up after big bang, yet it is the convergent gravitation field that is balancing the dynamic expansion of our expanding universe. Otherwise, it will be a fly away chaotic universe.

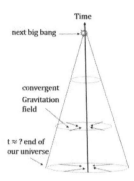

Figure 16.7 shows a convergent "de-creation" process due to gravitational force from a convergent PE developed by the degraded matters.

From creation to de-creation of our universes

Since creation of our universe can be understood from Hamiltonian standpoint, which was started from its divergent KE exploded, from which I had shown that de-creation of our universe must be from its convergent PE built by gravitation force from the degraded substances (e.g., dark matters and debris). Since only two dominant forces within our cosmological space; where KE explosion was derived mass annihilation, but total PE was due to Newtonian gravitation. Of which it fair to accept that the de-creating universe is due to convergent gravitational PE. Which creates the big bang scenario, but it is within a preexistent temporal (t > 0) space scenario to initialize the possible explosion. This is to rebut the conventional wisdom that our universe was created from a zero-summed empty space paradigm which is depicted in Figure 16.8. It shows a convergent de-creating universe after a divergent creation from the Hamiltonian standpoint. This paradigm may be one of the closest to the truth since science is approximated.

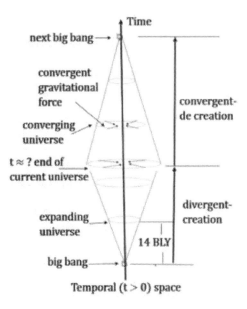

Figure 16.8 shows a convergent de-creation due to gravitational force after the end of a divergent creation of our universe.

A realizable Hamiltonian quantum machine

Since energy from mass annihilated is huge, this may be the reason why Schrödinger had opted to use Bohr's quantum leap into his machine. Nevertheless, like all classical mechanics, Bohr's atomic model was hypothesized within the same empty space platform for which his atomic model is not a physical realizable model. Instead, a realizable Bohr's model is hypothesized in Figure 16.9. In which quantum leap is a time and band limited wavelet (i.e., $\Delta E = h \, \Delta\lambda$), where h is a Planck's constant. From which Hamiltonian quantum operator takes place,

$$H(t): \Delta E(\Delta t) = h/(\Delta t), t > 0$$

which is known as Hamiltonian quantum wave equation, as given by,

$$H(t): [\Delta E(\Delta t) = h/(\Delta t)]\,\psi(t),\ t > 0$$

Since H(t): is constrained by (t > 0), the solution of ψ(t) is temporal, unlike Schrödinger's ψ(t) is time independent.

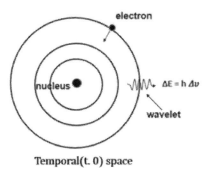

Temporal(t. 0) space

Figure 16.9 shows a realizable Bohr's atomic model situated within our time space.

A temporal (t > 0) Hamiltonian wave equation

When we incorporate a temporal eigenstate wave function
Ψ with a Hamiltonian operator, a Hamiltonian wave equation takes the form,

$$H(t)\psi(t): [-(\partial(PE))/\partial t = (\partial(KE))/\partial t]"\psi"(t),\ t > 0$$

In partial differential form we have

$$\{-m\,g\,\partial h(t)/\partial t = m\,v(t)d[\,v(t)/dt]\}\,\psi(t),\ t > 0$$

Where we see that as particle m falls it loses its PE but it is a gain of an its equivalent amount of KE. Notice that it is a physically realizable Hamiltonian wave equation in which the overall amount of energy is conserved. However, there are two cases to be discussed. Firstly, as applied H(t) ψ(t): to very large objects and the other is to small particles traveling at the speed of light.

Nevertheless, regardless how small a particle is, it cannot travel at speed of light. Which is the current constraint of our physics. In other words, when a particle reaches at speed of light, it must be self-annihilated into a package of EM wavelet radiations [i.e., $\Delta E(\Delta t)$]. From which the idea of particle-wave dynamic is equivalent to but not equals to a wave as a particle in motion or particle in motion is wave. For example, Bohr's quantum jump radiation is equivalent to a particle-wave dynamic but not due to particle annihilation. Thus, as a falling object's velocity reaches the speed of light, the object's PE is assumed of being converted its KE in light form [i.e., its $KE = (½) m c^2$]. But for falling object's velocity below the speed of light, its $KE = (½) m v^2$.

Back to the origin of our universe

Without time within a space even light cannot propagate, for this we have to agree that the origin of our universe has to be started (i.e., excitation t = 0) within a space of time (i.e., causality t > 0). Which is a greater temporal (t > 0) space that our universe embedded in [3]. As from Hamiltonian standpoint it is an isolated mass of m that must be situated within a temporal space as shown Figure 16.10. Since without the existence of time within a space, it is impossible for mass to convert to energy. If this is still
not a physical logic for you to accept, then I must have run out all the possible scenarios. Certainly, mathematic is not an option, since it is not physics. Any fundamental physicist treats mathematics as physics; it is Wrong. This is precisely why theoretical physics had been hijacked by mathematical oriented physicists. In view this hypothetical scenario has no PE, the only option is to convert its mass into energy is by means of mass annihilation. Fortunately, the hypothetical particle is embedded within a space having time. But mass annihilation needs an energy to trigger. A fair answer is that m has to be a gigantic mass created by a huge number of degraded substances, which were collected by their

gravitational forces that created a gigantic convergent gravitational PE to trigger the explosion from which without the convergent PE, it has no way to initiate the explosion. And this is the origin of our universe [3]. Although science is approximated, yet Einstein's 4-d spacetime continuum is so wrong.

[3] F.T.S. Yu, *"Origin of Temporal (t > 0) Universe"* CRC Press, Boca Raton, Fl (2020).

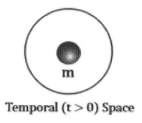

Temporal (t > 0) Space

Figure 16.10 shows a Hamiltonian particle embedded within a temporal (t > 0) space.

Conclusion

The essence of this presentation must be physical realizability for all fundamental principles (i.e., machines), since laws and theories cannot be virtual as is mathematics. In view of our current modern physics, it was hijacked by mathematical oriented theoretical physicists if not for decades. it must be for centuries. Since principles and laws were routed by mathematical analyses yet mathematics is not science. It is becoming more harmful than beneficial by scores of theorical physicists to keep promoting fundamental principles that are incorrect. Classical Hamiltonian is one of them.

My final question is this: would you want your students, children, and grandchildren to learn all of these fantasy principles? For example, a couple years before the covid pandemic I was invited to a Particle and Astro-Physics conference in Amsterdam. After I presented my talk, a young theoretical physicist approached me with a sincere piece of advice:

"Professor, why don't you learn our physics?"

"I am 87 years old," I replied. (This was back in 2019 when I was still a young man!) "Are you encouraging me to spend another 20 years to learn your kind of physics?"

I ask you, the reader: Why should anyone spend 20 years to learn something that is rooted in false assumptions?

Final remarks

Let me differentiate a distinctive difference between a theoretical physicist and a fundamental physicist. A theoretical physicist research how an object travels at speed of light from point A to Point B. But a fundamental physicist research for principles as to how a section of time Δt with an associated amount of energy ΔE is needed [i.e., $\Delta E(\Delta t)$] for a wavelet traveling at the speed of light from point A to Point B. And this is precisely why Theoretical physicists are so weird and Fundamental physicists are very realistic.

CHAPTER 17

Enigma of Eigen state wave function

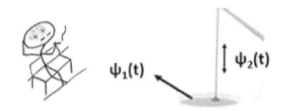

Examples, ψ(t) can be physically generated

Mathematics shows how ψ is implemented, but the material world does not tell us why ψ (t) works.

Enigma of eigen state of ψ(t)

Figure 17.1 depicts the myth of eigen state of ψ(t)

One of the mysteries of Hamiltonian mechanics must be "how" implementation of eigen state of ψ(t) in his mechanics but did not tell "why" he did it. Nevertheless, William R. Hamiltonian was not just a great scientist, or he was a clever fisherman as illustrated. In which a simple ripple of ψ(t) gave him a viable indicated expectation. For which this talk is all about the myth of eigen states of ψ(t), an unexpected player in the game of quantum physics.

From particle-wave dynamics to eigen state of ψ(t)

The aide of Louise de Broglie's particle-wave dynamics [1] means that particle in motion is "equivalent" to a wave dynamic. But we have frequently treated equivalence to as "equaled" to. This is precisely why scientists still cannot reconcile a photon as particle or a package of wavelet energy. Beside the equivalent wave dynamics, scientists virtually never mention why a particle in motion gives rise to wave? Secondly, what is the physical nature of the subspace that particle is embedded in that can create wave [i.e., ψ(t)]?

Nevertheless, why have these critical issues have never been raised? Firstly, our mathematical sciences were started from an empty space paradigm. Secondly since science needs mathematics, it must be the reason that our theoretical physics has been hijacked by math-oriented physicists. Thirdly, theorical scientists believed mathematics is science. For example, most physicists believe that it light moves spacetime but not time within space that moves light.

[1] E. MacKinnon, *"De Broglie's thesis a critical retrospective"*, American Journal of Physics, 44:1047-1055, 1976.

Photon is not a particle.

One of the most erroneous assumptions in modern physics must be that our universe was built by particles. Secondly most of them may not know that our universe is an energy conserved dynamics subspace. Since our time-space environment is one of the most important observable platforms, we knew that our universe is not just created by particles but is also filled with non-particle form substances. For examples such as permeability μ, permittivity ε, and others. And this is precisely why a package of wavelet energy has been wrongly treated as a particle like photon.

Since time is an independent continuous variable that coexists with space [i.e., temporal (t > 0)], every radiated wavelet of energy should not be treated as a particle like an electron. But it is a package of wavelet dynamic energy that travels at speed of light. For which a photon can be treated as a virtual particle (i.e., photon), which has momentum but no mass.

Physically realizable particle-wave dynamics

De Broglie's particle-wave dynamics was established within an empty space platform as depicted in Figure 17.2 (a). But for particle-wave dynamics to exist firstly, empty subspace must fill with non-particle form substances [e.g., fluid like substance (μ, ε)]. Secondly, the platform cannot be timeless, otherwise particle to wave dynamics cannot be created. From this, we see that empty space classical particle-wave dynamics is a false physically realizable model. Notice that this this one of the many classic examples (i.e., empty space paradigm).

Since it is time moves the wave but not the wave moves time of subspace, any physically realizable particle-wave dynamics scenario must be situated within a temporal (t > 0) subspace as depicted in Figure 17.2 (b). From which we see that, particle-wave dynamics exists within our time-space but not within an empty virtual mathematical space. Since empty space has no substance and no time to support a wave dynamic such as $\psi(t)$ which is the center topic of my presentation.

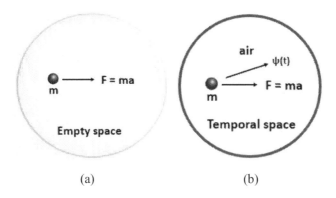

(a) (b)

Figure 17.2 (a) shows a Newtonian mechanics in motion within a virtual empty space paradigm, (b) shows the same Newtonian mechanics within a temporal (t > 0) space paradigm.

From Newtonian to Hamiltonian mechanics

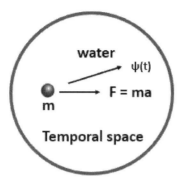

Figure 17.3 shows a Newtonian machine in motion. Without time within the subspace, the Newtonian machine cannot move.

Let me start from Newtonian Mechanics (i.e., $F = ma$) as depicted in Figure 17.3, where F is a force applying on a singularity mass assumption of m, a is its acceleration. Virtually our modern science was started from Newtonian mechanics within an empty space platform, from that it must be the reason why the foundation of our science, strictly speaking, is not physically realizable.

Even though, implication of the Newtonian changes with time, but without a platform that has time, $F = ma$ is a static mechanics. From which we see that, it is time of the space moves the machine [i.e., $F(t) = m\, dv(t)/dt$], but not $F = ma$ that moves time. From which we see that it is time of our universe (i.e., time coexist with space) that moves the Newtonian machine [i.e., $F(t) = m\, dv(t)/dt$], $t > 0$, as well moves us. In other words, we are living within a space that coexists with time, otherwise we have no place to go and no space to die.

Newtonian Physics: a momentum conserved engine

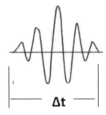

Figure 17.4 shows a package of energy wavelet.

Our universe is an energy conserved dynamic subspace, every physical mechanics is energy conserved (i.e., 1st law of thermodynamics). Thus, the momentum of a Newtonian machine within our time-space takes the form, $p = mv$ where m is assumed constant, but its velocity v is changing with time. Since momentum and energy are equivalent, an amount of momentum degraded equals to an equivalent amount of degraded energy left behind within the subspace in which the Newtonian is embedded. In view of particle-wave dynamics, the subspace of Newtonian machine embedded cannot be empty otherwise it cannot create a wave dynamic [e.g., $\psi(t)$], which is equal to the amount of degrading momentum Δp [i.e., $\Delta p(\Delta t)$]. Which is equivalent to a package energy wavelet shown in Figure 17.4. Since momentum and energy are equivalent, which is the equivalent amount of degraded energy $\Delta E(\Delta t)$. And this amount of degraded energy (or work done) produces a wave dynamic of $\psi(t)$. In which we see that within our universe, every change takes an amount of energy with an associated section of time [i.e., $\Delta E(\Delta t)$] to make it happen. Or a section of time with an associated amount of energy [i.e., $\Delta t(\Delta E)$], since time and its associated energy coexist.

Newtonian conservation operator

Since momentum is equivalent to energy, Newtonian conservation operator takes the form,

$$N(t): m\ v(t) = m\ \Delta v(t),\ t > 0$$

where m Δv(t) is an equivalent amount momentum to be degraded. Since subspace is not empty, that enabled mathematical oriented scientists to incorporate an eigen wave function of ψ(t) in the analysis. In other words, without wave generating medium, it is not possible to create ψ(t). By which Newtonian operator takes the form,

$$N(t)\ \psi(t): [mv(t) = m\ \Delta v(t)]\ \psi(t),\ t > 0$$

By way of ψ(t), scientists can use it to predict motion of Newtonian machine (i.e., F = ma), but with a limited result. As in contrast with an assertion of Max Born [2], ψ(t) can provide a full account of its particle dynamics. Nevertheless, the essence of incorporating ψ(t) is by way of indirectly determining its particle in motion, but it is by no means the whole account.

[2] M. Born, *The statistical interpretation of quantum mechanics*, Nobel Lecture, December 11, 1954

Hamiltonian: an energy conservative mechanics

One of the sophisticated machines in classical mechanics must the Hamiltonian mechanics as given by,

$$H: PE = KE$$

where PE and KE are particle's potential and kinetic energy. Of which it takes the form,

$$H(t): - m\, g\, d[h(t)]/dt = (½)\, m\, d[v(t)]^2/dt\, ,\ t > 0$$

where m is mass, g is a gravitational constant, and h(t) is the height of the particle. In which we see that PE decreases as h(t) decreases, but at the same time particle's KE increases. Since Hamiltonian operator is a radial operator, it takes the form,

$$H_r(t): - m\, g\, d[h(t)]/dt = (½)\, m\, d[v_r(t)]^2/dt\, ,\ t > 0$$

By way of cooperating an eigen wave factor ψr(t), we have the equation,

$$H_r(t)\psi_r(t): \{- m\, g\, d[h(t)]/dt = (½)\, m\, d[v_r(t)]^2/dt\, \}\, \psi_r(t)\, ,\ t > 0$$

In which by way of ψ(t), scientists can predict the dynamics behavior of Hamiltonian (i.e., H: PE = KE) but with little information provided by ψ(t) as in contrast with the common believe that ψ(t) has a full account of its equivalent particle dynamics.

Legacy of "classical" Hamiltonian mechanics

So far, a successful legacy of classical Hamiltonian must be Schrödinger's quantum mechanics. But Schrödinger's theory is so wrong since classical Hamiltonian is real. Although the success of Schrödinger quantum mechanics is due to adaptation of Bohr's quantum leap, that why it took the name as quantum mechanics. But it is not proof that his quantum machine is correct, since Schrödinger wave equation [i.e., ψ(t)] is not a time limited temporal (t > 0) wave equation. By way, adopting Bohr's quantum leap hypothesis is not proof his quantum theory is right, since Bohr's model is not a physically realizable model. Firstly, Bohr's atomic model is situated within an empty space platform, as does all the classical mechanics. Secondly, Bohr 's quantum leap hypothesis is not a time-limited physically realizable wavelet. Even though a bunch of reasons can show Schrödinger's quantum mechanics is not physically realizable, yet his fundamental principle has been accepted by world physical community for decades or near a century. This is precisely why our modern physics is so wrong since we were too dependent on mathematics. For example, how many physicists will agree that light travels with time but not change time of the space. If anyone can figure it out why, then he may know why modern physicist is so wrong.

Essence of ψ(t)

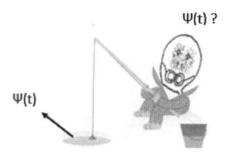

Figure 17.5 shows a ripple of ψ(t) giving rise to a viable prediction..

A viable diagram is worth more than pages of descriptions. In Figure 17.5 , we see that a ripple of ψ(t) gives rise to a valuable prediction for an icy fisherman. But it is by no means a full account of particle dynamics (i.e., particle-wave duality) as provided by a ripple of ψ(t). Since ψ(t) is caused by particle in motion, but they are somewhat equivalent but not equaled. This is exactly what Hamiltonian has done to his mechanics. By way of ψ(t), he can predict with high success of past experiences, even though by way of ψ(t) does not represent a full account of actual particle in motion. In which we have seen the essence of ψ(t), but not as the conventional assertion of Max Born. Where ψ(t) does not represents a full account of the particle dynamics.

A last ditch of ψ(t)

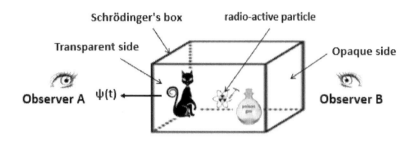

Figure 17.6 shows a paradox of Schrödinger's cat.

Since Schrödinger's half-life cat has taken a toll on our modern science, yet a simple illustration can resolve decades of mathematical physicists' mystery. For example, Schrödinger's experiment is performed within one-sided transparent box as depicted in the figure. Since observer B cannot observe ψ(t), he must wait until someone opens the box for the great superposition principle to collapse. Yet, observer A precisely knows the cat is either dead or alive before opening the box. I think this simple illustration should clear-up years of theoretical physicists' puzzle, instead of being buried within the forest of mathematics.

A note on quantum world

In view of Schrödinger's cat hypothesis, it is not just a physically non-realizable postulation but a nonsense postulation that had led to a worldwide quantum conspiracy. The damages have had been done; it is not just billion but trillion of taxpayer's money had been used to a non-existent hypothesis. Otherwise, it could have been used to a more constructive science. Or in part, it can be allocated to save our planet and humanity.

Furthermore, no matter how beautiful mathematics is, it cannot alleviate a non-realizable hypothesis. For which it has been debated for almost a century from a mathematical standpoint but has not been looked at from the physical hypothesis standpoint.

Physical reliability of $\psi(t)$

With reference to preceding Schrödinger's cat, it turns out to be an excellent example of a great principle that touches millions of innocent followers [i.e., pied piper effect]. In which we have seen it is not just billions, but trillions of taxpayers' money invested in a principle which does not even exist. Regardless of whether $\psi(t)$ is a sound wave or a light wave, every observation takes a section of time to observe and another section of time to reach the observer. For which, I wonder why we are still debating a principle which does not even exist within our universe? Furthermore, this excellent case shows that a wrong physically realizable hypothesis has taken a toll of decades of irony debates. In other words, the paradox of Schrödinger's cat is not lying by incompleteness of mathematics, but the physical realizable paradigm.

If one understands that it is time of our space moves wave (or velocity), but not wave (or velocity) moves time, then one should know why theoretical physics is so wrong. For example, we have,

$$\psi(t) = \cos(\omega t), t \geq 0$$

$$V(t), t \geq 0$$

that are a set temporal (t > 0) equations as depicted in Figure 17.7 where ω is radian frequency. From which we see that the pace of time (i.e., our universe) moves the wave, but not wave or velocity changes the pace of time. In other words, waves or particles cannot move without the existent of time.

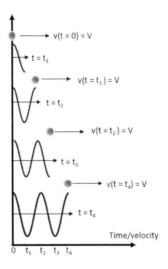

Figure 17.7 shows that wave and velocity cannot alter the pace of time.

Conclusion

I have shown how the eigen state wave function ψ(t) is implemented by math-oriented physicists, but they did not tell us why ψ(t) works. Nevertheless, incorporating ψ(t) with the analysis is rooted by de Broglie's particle-wave dynamics. In words, by way of ψ(t) we may predict particles in motion with some success. Since Hamiltonian mechanics was built on an empty space platform, implementation of ψ(t) suggests that platform is not empty, otherwise ψ(t) has no way to propagate. And this is exactly how our analytical physics was developed. ψ(t) is valuable information but is very limited. This contrasts with legendary Max Born's assertion, that the eigen wave function $\psi(t)$ provides a full account of particle dynamics.

Nevertheless, implementation of $\psi(t)$ is a way of understanding the dynamics of particle in motion that produces $\psi(t)$, that normally is unable to detect.

A message to math-oriented physicists

Two of the most well-regarded modern physicists of all time are the inventor of relativity theory and the creator of quantum mechanics. Yet they are also two of the most prestigious conspirators in modern science's history. But the fault is due to us, since we knew modern physics is so weird, but we keep promoting it. Maybe we are a flock blinded and accustomed to follow. For example, Richard Feynman had warned us, "after we had learned quantum mechanics, we simply do not understand quantum mechanics". Yet, after all those years, we did not even try to find out why? And this is our own fault and irresponsibility.

Since mathematics is not science, if we keep promoting those mathematical fantasy sciences it is more harm than beneficial. Personally, I am sympathetic to those who spent decades or even lifetime on this non-existent physics. But sympathy cannot shadow the disappointment since they have not shown any attempt to correct it.

CHAPTER 18

Why mathematics is not science

Mathematics is a tool to discover science but itself alone is not science.

$\int f(x) \, \pi \, \Omega \, dx \neq mc^2$

Mathematics is not science, but science is mathematics?

Fundamental science was lost in the woods of virtual mathematics, but I found it in a temporal (t > 0) space promised land.

Why mathematics is not science?

Mathematics is created by humans; it does not need to obey the laws of nature as science does. For which mathematics does not physically exist but is virtual.

Science is a part of nature; it must obey the laws of nature. Although science is approximated, it cannot be created. And even though technological sciences create, they must, nonetheless, obey the fundamental principles of science. Thus, all the fundamental principles must be discovered within the laws of nature.

In contrast, mathematics permits paradoxes.

What is mathematics?

From scientist's standpoint, mathematics is a symbolic language, like our written or spoken languages. But it does not need to obey the laws of nature since they were virtually created by humans but not physically discovered. For example, a set of equations as given by,

$$A = B, A - B = 0, \text{ and } 0 = B - A$$

From mathematical point of view, they are the same equation. Similarly, a given a principle of,

$$f(t) = \cos(\omega t)$$

where ω is the radiant frequency and t is a time variable. Notice that this assumed equation has no time limitation on it; it can be used within our time-space.

The reason mathematics does not need to obey laws of nature since it is not a part of nature. Unlike science, which is part of physical entity which cannot escape the laws of nature. For this we see that, it is terribly wrong for math-oriented physicists to treat mathematic as science.

What is science?

Since science is a part of nature; it must obey the laws of nature. So, what are the laws of nature? Since our universe is an energy conservation dynamic subspace, law of time, law of energy conservation, law of thermodynamics, law of entropy, are a few examples. From which any fundamental principle cannot simply violate any of these laws. Let me further stress that, these laws of nature were discovered but not created.

Interpretation of an equation from physical reality standpoint, is very different from mathematical viewpoint. For example, $A = B$ is interpreted a conservative equation in physical standpoint, while $A - B = 0$, or $0 = B - A$ are zero-summed equations, as from energy conservation standpoint.

In which, $f(t) = \cos(\omega t)$ is not obeying law of temporal $(t > 0)$; it is not a physically realizable equation that can be applied within our time-space.

Although mathematics is not science, science often uses mathematics since scientific principle can be easily described by simple mathematical equations. Mathematics is not science, even if science is not easily dissociated from mathematics.

What are the laws of nature?

To understand the laws of nature, we must first know how our universe was created. Since it is impossible to understand exactly our universe was created, but there is a closest to truth model by current reliable knowledge of physics that can be used in the name physical realizable paradigm [1].It is however a big mistake to use a created spacetime continuum that violate the laws of nature [2].

So, what are all the existed laws of nature that we already have? Those must be law of temporal (t > 0) (i.e., time), law of energy conservation (i.e., 1st law of thermodynamics), law of entropy (i.e., 2nd law of thermodynamics), and others.

For example, Einstein's 4-d spacetime continuum is a created model that violates all the laws of nature. This shows that any principle developed from Einstein's spacetime model is doomed to be false, even though its mathematical presentation is correct. From which we see that it is not how rigorous and correct mathematics you used, it is the right paradigm used that will give us a physically correct principle.

[1] F.T.S. Yu, *"The Nature of Temporal (t > 0) Science: A physically Realizable Principle"* CRC Press, Boca Raton, Fl (2022).

[2] A. Einstein, *Relativity, the Special and General Theory*, Crown Publishers, New York, 1961.

Law of temporal (t > 0)

Since our universe is an energy conserved dynamic subspace, in which it obeys the law of temporal (t > 0) [i.e., casualty]. For example, take mass to energy equivalent equation as given by,

$$\Delta E = \tfrac{1}{2} (\Delta m) c^2$$

which shows that an amount of energy ΔE increases is coming from an equivalent amount of mass Δm annihilated. But it is a timeless equation, without the intervention of time, it has no way for the conversion. For which Δm must already exist within a space with time. In this one of the examples shows that physical substance cannot be existed within an empty [(i.e., timeless (t = 0)] space period. And this is one of the most important law in nature since our universe (i.e., our living spacetime) cannot be empty. Call it the temporal exclusive principle—one which needs an expenditure of Δt (i.e., a section of time) to make the mass to energy conversion happen (i.e., a law of nature). For this, without imposition by temporal (t > 0), just from mathematical interpretation, can be justified that mathematics is science. Furthermore, from mass and energy equivalent equation, we cannot even tell it is a mass to energy or energy to mass conversion, unless the equation is imposed, but law of temporal (t > 0) is the one that cannot be violated.

Application of mathematics to science

Since science needs simple mathematical descriptions to facilitate the complexity of science, in view of limited ancient civilizations, it would be a surprise if Chinese characters were selected to develop mathematics. For which Arabic, Greek, as well English alphabets were used in mathematics. As from a scientist's point of view, mathematics is a symbolic language. This is precisely why mathematics were mostly developed by westerners. And this one of the reasons, why science needs mathematics but mathematics does not need science.

Fault of mathematical physics

Nevertheless, fault of modern physics is rooted by the created zero-summed spacetime paradigm of Einstein. This must be the reason why his spacetime paradigm have never been challenged. Although you may not accept our universe is an energy conservation dynamic subspace, but it is surely not a zero-summed time-space of Einstein. For example, no matter how fast that you can travel, you would never be able to see your own grave.

Since practically all the fundamental principles were created by math-oriented physicists in which they regarded mathematics as science. This trend has led us in the woods of mathematics, if not for decades must be over a century. Since the fault is due to math-oriented physicists, since they knew our modern science is so irrational, yet they have opted to promote it.

Nevertheless, every discovery depends on a physical realizable platform (i.e., laws of nature). In the following I will show by imposing laws of nature, realizable science could be found. Which means that mathematics cannot be science by itself, since it does not need to obey laws of nature.

What is a zero-summed universe?

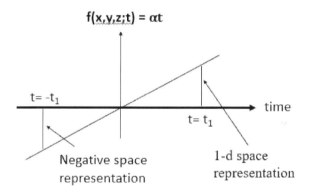

Figure 18.1 shows a one-d spacetime within a zero-summed universe model.

Although knowing that modern physics is irrational, math-oriented physicists continuingly relied on mathematics for solutions. Figure 18.1 depicts a zero-summed spacetime model from a mathematics viewpoint where $f(x, y, z; t) = \alpha t$, represents a 1-d spacetime as a function of time t and α is an arbitrary constant.

In view of the preceding diagram, we see that space changes with time, its symmetric negative space [i.e., $f(-x,-y,-z; -t)$] started from $t \rightarrow -\infty$, to t = 0, and then to positive space [i.e., f (x, y, z; t)] all the way toward at t → ∞. Nevertheless, if we accepted our universe model as it is, in principle all the spacetimes in the time domain are existed, as from a blindly mathematical interpretation. This is virtually the same to a well-accepted model, that anti-matters are existed within our current universe, since all the past spacetimes are still there, yet we are in the current spacetime. This is exactly the reason, without imposition law of nature, that mathematics alone can wrongly interpret a principle.

An energy conservative universe

Firstly, let me take a mass-energy equivalent equation as given by,

$$E = \tfrac{1}{2} mc^2$$

which is a non-zero summed equation. Similarly, it can be written by,

$$\Delta E = \tfrac{1}{2} (\Delta m) c^2$$

which shows that an amount of energy ΔE increase is due to an equivalent amount of mass Δm annihilated. But it needs an expenditure of Δt (i.e., a section of time) to make it happen (i.e., a law of nature). From which without an imposition of time, a mathematical expression can hardy justify as science.

Yet as from mathematical standpoint this equation can be equivalently written by a set zero-summed forms,

$$E - \tfrac{1}{2} mc^2 = 0, \quad 0 = \tfrac{1}{2} mc^2 - E$$

These show there is a symmetry exchangeable property among energy and mass. But this formulation has misled us to interpret that our universe was created by a zero-summed empty space paradigm, instead by an energy conservative dynamic subspace [3]. Since empty and non-empty are mutually exclusive, our universe cannot allow any empty timeless subspace to exist.

[3] F.T.S. Yu, *"Origin of Temporal (t > 0) Universe"* CRC Press, Boca Raton, Fl (2020).

Mathematics is not science, firstly, we must understand how our universe was created. By which a question for you is that what is the distance between yourself and yesterday of you? If you knew the answer, you may have the idea of how of our living time space was created. From which you may see mathematics itself is a tool which cannot be treated as science by itself alone. And this exactly what math-oriented physicists have done to our fundamental physics since they treat mathematics as science.

As can be seen, if one imposes law of temporal (t > 0), one may find ways to tell if a principle is either science or not. For example, by imposing law of temporal (i.e., t > 0) on Cos (ω t) as a physically realizable principle. On which a [Cos (ω t), t > 0] equation is a temporal principle. Similarly, without the imposition law of temporal (t > 0) on Schrödinger's eigen wave equation, his superposition principle has created a worldwide quantum conspiracy. Since quantum scientists have literally treated his non-existed superposition principle as physically real.

Nevertheless, mathematics itself is a usable tool but alone by itself cannot be a science. From which we see that, no matter how rigorous mathematics is, it should not be treated as science, but a viable tool for mathematical physicists.

Dirac anti matter hypothesis

Another example, take the Dirac equation as given by[4],

$$(i\not{\partial} - m)\psi(t) = 0$$

Since it is a zero-summed equation, it indicates that anti-energy and anti-mass exist within our time-space. But those anti-matters or anti energy were supposed to have exited in the negative time of the universe. Then how can Dirac justify simultaneously of anti-matters within our current time-space? For which we see that, without the imposition by law of temporal (t > 0), any solution developed based on Dirac equation cannot be physically legitimized. Nevertheless, if Dirac equation takes a non-zero summed form,

$$(i\not{\partial} - m)\psi(t)$$

Then his anti-matter postulation may not be hypothesized. From which, without the imposition laws temporal (t > 0), even using rigorous mathematics cannot save his equation. From which we see that it is the physical realizable platform that warrants a physical realizable equation, but not by mathematics alone.

[4] A. Hirschfeld, *The Supersymmetric Dirac Equation*, Imperial College Press, London (2011).

Feynman's quantum electrodynamics

Regardless of the complexity of Feynman's QED analyses, all his QED solutions are doomed to be false and fictitious as mathematics are. If one imposes law of temporal (i.e., t > 0), one can easily show that all his QED solutions are virtual as mathematics are and even fake. This is a very interesting topic that I have left in part of you to fill in.

Nevertheless, the fault is mostly due to us, since we are accustomed to being good followers. Although he had warned us as I quote; "after we had learned quantum mechanics, we simply do not understand quantum mechanics". In which we see that Feynman himself was also one the faithful followers.

Yet, after all those years, we are still relying on purely mathematics to answer our science needs. This is like searching in the woods of mathematics for physical realizable principle, but physical realizable principles are waiting for us to discover in land of temporal (t > 0) universe. This is precisely what math-oriented physicists have misled us for decades or over a century. But the irony is that without being equipped with mathematics, physicists have no tool for them to discover.

A paradox of a great theory

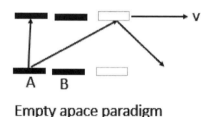

Empty apace paradigm

Figure 18.2 shows a classic textbook mathematical physics example.

Nevertheless, it is mathematics that Einstein used to discover his legendary relativity theory as depicted in the Fig. 18.2 . In which we see that without mathematics, it is impossible for him to develop his theory.

In which two sets of light beams are synchronously bouncing back and forth between two sets of mirrors; one set is stationary, and the other is moving away at a constant speed of v from a stationary set. If we equip two clocks A and B respectively within each set, mathematically we have found clock A runs faster than clock B, if we stay on the moving platform of B. On other hand, if we stay with the stationary platform A, we will discover clock B moves slower than clock A. Notice that this is a classic textbook material, but we have shown Einstein's relativity theory is dead wrong, but why do we keep promoting it?

On a pure mathematical platform

In view of the depicted Figure 18.3 , it is very likely math-oriented physicists would agree that the apple is falling. But inadvertently they have violated their own mathematical principle since mathematically speaking the apple is either falling or our planet is moving up, as based on Einstein's relativity postulation.

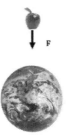

Figure 18.3 shows an apple falling from an apple tree.

So, my question is this, would a mathematical oriented physicist stick with the theory that he has learned, or opted toward physical realizable science? From which we see that without any restrain by laws of nature [e.g., temporal ($t > 0$)] just depending on mathematics alone is not adequate to resolve the reality issue of science. For a scientist, mathematics is a tool that should be used for new discovery, but again mathematics alone cannot alleviate the fundamental problem of physics. In view the work of EPR [4] they blamed it as mathematically incomplete, instead of a violation of law of temporal ($t > 0$).

[4] A. Einstein, B. Podolsky, and N. Rosen, Can *Quantum Mechanical Description of Physical Reality be Considered Complete?*, Physical Review 47 777(1935).

Mathematics moves a wave?

An electro-magnetic wave propagation is given by,

$$\nabla \cdot S = -(1/2) \, d(\varepsilon E^2 + \mu H^2) \, dt$$

In which mathematically assumes a radiating EM wave travelling within a space of electro-magnetic medium (i.e., μ, ε). This equation shows that an EM wave is moving within a space filled with permittivity substance. Yet, my question is: does the wave move with the time of the medium, or the time of the space move the wave?

Since mathematically (i.e., the equation) indicates that it can be both ways, yet physical reality has only one choice. However, if we introduce law of temporal ($t > 0$) with the consideration, the answer becomes apparent. This show that mathematics alone cannot fulfill the name of science. Yet it is a vital device, without it theoretical physicists go nowhere. But pure mathematics cannot be accepted as science as from a scientist's standpoint.

Conclusion

In conclusion I have made a distinction between fundamental science and mathematics. Fundamental sciences are mean to discover the laws of nature, which cannot be created by humans, while mathematics is created by humans. From which it is wrong to treat mathematics as science, as from a scientist's standpoint. Yet science needs mathematics to describe its fundamental law, and this is exactly why science was hijacker by math-oriented physicists, for decades or even centuries.

In view of the currently dominated mathematical physics, we see that catastrophic damages have been done. But this is about time for us look for viable physical realizable sciences. Otherwise, we will forever be trapped within a fantasy mathematical physics which will be more damaging than beneficial to our science, and to our humanity. Since trillions of research revenues have been wasted to principles that cannot even exist within our spacetime. To those who spent decades or even lifetime on this non-existent physics, my sympathy cannot shadow my disappointment. Because, they have not even shown any attempt to correct it, instead openly keep promoting it.

CHAPTER 19

Origin of our science

Empty space is not a physically realizable paradigm.

A mathematical simulated blackhole

Legacy of a fantasy scientist

One of the most intriguing questions in life must be where science comes from. Since we are a part of science, how our science developed from must be a legitimate question. At the dawn of our science, it must be started by a schematic diagram with a script of mathematics on a piece of paper or papyrus. That must be the subspace that we used to develop all our sciences.

The essence of this presentation is to show that the emptiness background of that piece of paper does not emulate a physically realizable subspace, since our living space is compacted with substances and has time. That piece of paper represents an absolute emptiness background, which has no substance and no time. This is the emptiness subspace that we have used to developed all of our sciences. This is the reason that piece of paper made our sciences, classical and modern alike, being as virtual and fictitious as mathematics.

Where did our science start from?

Mathematics is created and science is to be discovered, but there is a correlation between them. Mathematics alone is not science, but there is a need to develop science. From a scientist standpoint mathematics is a symbolic language which is useful to describe complicated science. This must be the reason why science needs mathematics, but mathematics does not need science.

All of our sciences were started from a diagram (i.e., a 2-d sketch) on a piece of paper, with arrays of symbolic equations to describe science (i.e., 1-d presentation), as depicted in Figure 19. Symbolic mathematics is one of the efficient languages to facilitate the diagram, which is known as "science". From this we see that science is mathematics.

We have assumed that the underneath subspace is empty, from which all our sciences were developed. The fundamental principles that include our modern physics were developed on the same empty subspace. This is the reason why our science is a "deterministic"(i.e., exact) science, which already violates the law of entropy.

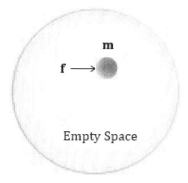

Figure 19.1 shows schematically f is applying a force on a stationary mass of m. For which we have f = ma, where a is its acceleration.

Believe me or not

Every equation in our physics or engineering textbook is deterministic. Strictly speaking they are not physically realizable since every science change with time. For instance, a simple ohm's law is deterministic.

Nevertheless, the foundation of our science was built on an empty paradigm, for which all the principles and laws are anticipated to be as exact and virtual as mathematics. This is precisely the reason why all our sciences are deterministic, yet deterministic is not physically possible. In other words, any science developed on this virtual empty subspace is doomed to be as virtual as mathematics is, and therefore wrong.

Yet, for centuries we have developed countless laws and principles from the empty space, not knowing it is not a physically realizable paradigm. Modern physics demanded instantaneous (i.e., time), and simultaneous (i.e., space) connections, as well as the changing pace of time. I found that the foundation of our science was built on a virtual mathematical empty space, which is not physically real.

Why does classical science work?

It appears that if the foundation of our classical physics is not real, why do the principles work in practice? The answer is that those working principles that work do not violate the laws of nature. Secondly, those applicable sciences are approximated. The deterministic or exact solution will not work, because everything within our time-space changes with time. We see that our intervention of bending the rule of science (i.e., the deterministic or exact principle) makes it work approximately.

Since Einstein's relativistic theories are deterministic, we see they violated the law of causality [1], the temporal ($t > 0$), the law of entropy, as well as others. Furthermore, if one picks up a principle from any published physics book, there is a good chance that principle has violated the law of time (i.e., $t > 0$ or causality principle). This is precisely the reason why the foundation of our
modern physics does not physically exist since it is developed on an empty timeless space platform.

Even though causality was a part of science, math-oriented physicists excluded it from consideration. Therefore, it is very unlikely for them to develop modern science. But it is mathematics, which has been accepted as science by most physicists.

[1] M. Bunge, *Causality: the place of the causal principle in modern science*. Harvard University Press, Cambridge (1959).

Nevertheless, since every instant is the absolute physical reality in time (i.e., t = 0), it must be the reality that the causality principle prevents the instant response of excitation. I am certain the discovery of causality in classical physics was not considered. Otherwise, modern physics might not have developed. Nonetheless, by experiencing the reality in time, we have found everything within our universe has a life, which includes the life of our universe. In other words, everything that changes within our universe has a price tag, an amount of energy ΔE and a section of time Δt [i.e., $\Delta E (\Delta t)$] and it is not free. Therefore, we see that every mathematical principle cannot be a physical realizable principle since mathematics alone cannot be science. But science is mathematics. Which is precisely the dilemma that we encounter in science.

Modern physics was built on the same emptiness subspace.

Regardless, modern physics is the extension or legacy of classical physics, since it used the same empty space platform, for which all the fundamental laws and principles in modern physics have suffered with the same non-realizable solution (e.g., deterministic). It is the underneath platform that determines the reality of the principles, but not the mathematics. This must be the reason why modern physics is so weird and so wrong [3].

For decades or over a century, we have encountered illogical solutions from modern physics, but mathematics has over-shadowed the reality. This must be due to the fact that modern physics was developed by a group of math-oriented physicists. For example, if there is a problem, they blame mathematics [4]. The empty space paradigm has never been a place to look. Since our science was created by a group of math-oriented scientists, this must be the reason why mathematics is always to blame. I found that most math-oriented physicists do not even know how our science got started. Apparently, it was started on a virtual mathematical empty space, and every math-oriented physicist knows what an empty space can do to our science.

[3] F. T. S. Yu, *"What is "Wrong" with Current Theoretical Physicists?", Advances in Quantum Communication and Information.* Chapter 9, p 123-143, IntechOpen, London (2020).

[4] A. Einstein, B. Podolsky, and N. Rosen, Can *Quantum Mechanical Description of Physical Reality be Considered Complete?*, Physical Review 47 777(1935).

What triggers the discovery of temporal (t > 0) space?

Traditionally, if things work well, we follow on the same path that gives a better chance of success. Although classical physics has never encountered any major crises in the past, it is because we have already accepted that science is approximated. This must be the reason those modern physics' principles triggered my interest. I have shown that it is the underneath platform that decides the physical reality, not the mathematics. I found it is the empty space paradigm, new and old, that causes the non-physical realizable problem. Since mathematics alone cannot change the time dependent issue of science, I found it is the underneath subspace that gives the causality or temporal (t > 0) aspect of physics [5]. If we are depending on mathematics alone, we cannot alter the physical reality of science.

[5] F.T.S. Yu, *Origin of Temporal (t > 0) Universe*, CRC Press, New York, (2019)

A viable example tells it all

Before wrapping up, let me show an example to epitomize my issue. Let me take de Broglie's wave mechanics as depicted in Fig. 19.2 (a). Firstly, we see that a moving particle is situated in an empty space paradigm. Similarly, an identical scenario, but situated within a temporal (t > 0) space is shown by Fig. 19.2 (b). Since subspace of Fig. 19.2 (a) is empty, it is impossible for a moving particle to create a wave. Therefore, empty space is not a viable physically realizable model. However, if the same hypothesis is submerged in a temporal (t > 0) subspace of Fig. 19.2(b), a dynamic wave of $\psi(\sigma;t)$ or $\psi[\sigma(t)]$ can be created, where σ represents a space coordinate system.

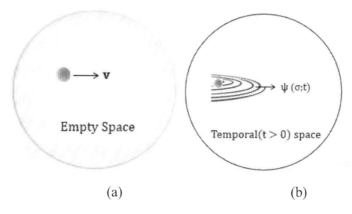

(a)   (b)

Figure 19.2 (a) shows a de Broglie particle-wave dynamics within an empty space, b) shows a de Broglie particle-wave within a temporal (t > 0) space scenario.

The fact is that the foundation of our science was created on an empty virtual subspace; this is precisely the reason why all of our sciences (including our modern physics) are not physically realizable. Science cannot be absolutely deterministic. However, our modern physics want us to be deterministic, which is against the laws of nature.

Conclusions

For centuries we have benefited from those timeless laws and principles. We have never encountered any major set-back, not until modern physics came along (e.g., Einstein's relativity and others). This must be due to the demand for instantaneity (i.e., time), simultaneity (i.e., space) and changing the pace of time (i.e., law of time) that caught my attention.

Since modern physics has given us weird and irrational solutions, this must be due to math-oriented physicists that prevented us from finding them. Instead, they come-up with an irresponsible reason to coverup them up. For example, they created a fictitious micro and macro space scenario to legitimize their irrational results. Modern physics is rooted with the same empty space paradigm that apparently violates the first law of nature [i.e., causality or law of time (i.e., $t > 0$)].

Although the causality principle had been a part of classical physics, but math-oriented physicists have apparently ignored it. Otherwise, modern physics would not have developed. For instance, I quote legendary Richard Feynman on quantum physics "After we have learned quantum mechanics, nobody is really understanding quantum mechanics." Yet he himself followed the principle. This must be one of the reasons that we have not even tried to understand quantum mechanics. Since Feynman could not understand, why should we waste our time.

CHAPTER 20

Origin of de Broglie wave dynamics

Broglie wave dynamics is virtual

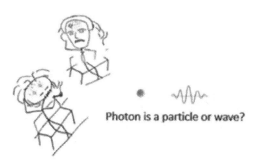

He must be one of greatest virtual physicists.

One brilliant idea in mathematical physics must be credited to Louse de Broglie's particle-wave dynamics. In which a package of wavelet energy initiated by means of a particle's dynamics as depicted. For which his wave mechanics is a virtual mathematical deterministic principle as all our mathematical sciences are.

Origin of de Broglie's wave dynamics

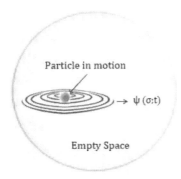

Figure 20.1 shows a particle in motion creates a wave dynamic of $\psi[\sigma(t)]$, but within an empty space. Notice that this empty space paradigm is the origin of all our sciences were developed.

One brilliant idea in mathematical physics must be credited to Louse de Broglie's particle-wave dynamics [1], in which a package of wavelet energy is created by means of a particle's dynamics as depicted in Figure 20.1. Since empty space paradigm is the origin of all our mathematical sciences, which is precisely why all our sciences, new and old, are deterministic (i.e., predictable with absolute certainty). By which de Broglie particle-wave dynamics is a virtual mathematical principle, deterministic as all our mathematical physics. Since mathematics is not science, but science is mathematics [2]. Without the imposition of laws of nature [e.g., law of temporal ($t > 0$)], solution as obtained from de Broglie wave mechanics is virtual as mathematics is. For example, one of erroneous interpretation from de Broglie's particle- wave dynamics must be the interpretation a package of wavelet energy {i.e., $\psi[\sigma(t)]$} is equivalent to a particle like "photon" in motion, but photon is not a particle.

Besides, de Broglie wave dynamics is a mathematical principle; its background subspace is empty. As from physical reality standpoint, how can he justify that particle in motion can create wave. From which we see that his particle-wave dynamics is a deterministic timeless principle, which is rooted by empty space paradigm.

[1] E. MacKinnon, *"De Broglie's thesis a critical retrospective"*, American Journal of Physics, 44:1047-1055,1976.

[2] F.T.S. YU, *"Introduction to Physical Realizable Physics"*, Page Publishing, Meadville, PA (2023).

What a particle-wave dynamic means

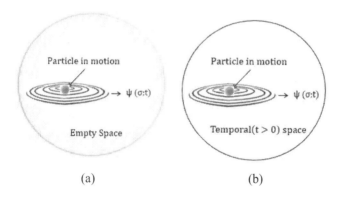

(a) (b)

Figure 20.2 (b) shows a particle in motion creates a wave dynamics of $\psi[\sigma(t)]$ within a temporal (t > 0) space. Notice that temporal space paradigm is a physical realizable paradigm.

Within our time-space, entropy increases naturally with time. For which a created wave $\psi[\sigma(t)]$ loses its dynamic relationship (i.e., correlation) with the particle that induced the wave. In which we see that, it is more difficult to determine their relationship by means of its wave dynamics $\psi[\sigma(t)]$ as time move further away. Evidently, as from de Broglie's wave dynamics {i.e., $\psi[\sigma(t)]$} standpoint their relationship is deterministic. For which his particle-wave dynamics is a virtual mathematical principle, since it was developed on a virtual empty space as again depicted in Figure 20.2(a). Thus, any solution comes out by this de Broglie's wave dynamics principle will be deterministic and virtual as mathematics is.

.....However, if de Broglie's particle-wave dynamics hypothesis is situated within a temporal (t > 0) space as depicted in Figure 20.2 (b), its induced wave dynamics $\psi[\sigma(t)]$ will change with distance (i.e., time). In other words, its relationship with respect to the moving particle decreases with time distance. In other words, further away the wave dynamics $\psi[\sigma(t)]$ is, away from the dynamic particle, the more uncorrelated (i.e., unpredictable). For which it is wave $\psi[\sigma(t)]$ changes with time. But in contrast with the classical de Broglie's principle, wave dynamics remains the same regardless of distance, which is a time invariant or independent wave function.

A physically realizable de Broglie's wave dynamic

Nevertheless, if de Broglie's particle-wave dynamics hypothesis is situated within a temporal (t > 0) subspace paradigm as depicted in Figure 20.3 any solution comes out from this model would be very likely physically realizable or obeyed law of time [i.e., temporal (t > 0)]. Since within our temporal universe, every subspace is compacted with physical substances that changes with time [i.e., temporal (t > 0)] it filled with non-particle like material such as electro-magnetic medium (i.e., permeability μ and permittivity ε). From which a package of electro-magnetic wavelet {i.e., $\psi[\sigma(t)]$} can be created. In fact, this is the "sufficient" condition for de Broglie's particle-wave dynamics to realize. This is the reason his principle is virtual as a mathematical principle.

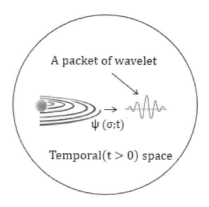

Figure 20.3 shows photon is not particle.

Since particle in motion is a "necessary" condition to excite a physical EM wave of $\psi[\sigma(t)]$, from which we see that a package of energy wavelet is not a particle like photon as depicted in Figure 20.3. But we had mistakenly interpreted a package of wavelet as a photon. Which is one of a serious mistake to equate a wavelet of $\psi[\sigma(t)]$ travels at speed of light to a particle like photon, as a "duality" of its particle dynamics. For examples, a vocal cord is not a voice, or a string is not a wave.

[3] F.T.S. Yu, *"Origin of Temporal (t > 0) Universe"* CRC Press, Boca Raton, Fl (2020).

What de Broglie's wave dynamics has done to science?

Since Broglie's wave dynamics is a well thought purely mathematical principle, but it is not corresponding to any physical reality within our temporal (t > 0) space. Like all our sciences are, because the origin of our mathematical science was started on an empty mathematical virtual space. This is precisely why all our sciences, new and old, are deterministic. For instance, pick any principle that you can find in any physics or engineering textbooks, you will find it is either timeless or deterministic. Yet, every physical existent principle within our universe changes with time and approximated. But this is the deterministic or timeless principles that our universe cannot provide because our universe is a dynamic energy conservation expanding subspace.

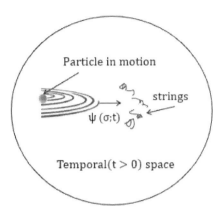

Figure 20.4 shows a set of mathematical equivalent springs which is not a by-product of particle dynamics to replace $\psi[\sigma(t)]$.

Nevertheless, de Broglie's wave dynamics is one of those principles developed on the same empty space platform; which is anticipated that de Broglie's wave mechanics is a virtual mathematical principle, but not physical realizable law. Yet it must be the mathematical nature de Broglie's wave mechanics that has been wrong treated within our time-space. For example, String Theory [4] is one the legacies of de Broglie's wave principle, but mathematical equivalent is not the same as to a physical reality equivalent. For which String Theory uses a set of fix-ended strings to emulate de Broglie's wave dynamics. Since a set of equivalent strings is not a by-product of particle dynamics, it does not exist a relationship with respect to its particle's dynamics. In which we see the string theory is a pure mathematical equivalent with particle's dynamics, but not a physical equivalent principle. In short but a purely mathematic gamic in mathematical physics.

[4].. B.. Zwiebach, *A First Course in String Theory* 2nd ed. Cambridge U. Press (2009

What m ψ[σ(t)] really means

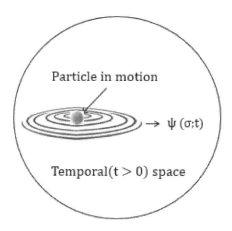

Figure 20.5 shows a particle to wave dynamics where {ψ[σ(t)], (t > 0)} moves away from particle motion.

There is an intimate relationship between particle dynamics {i.e., m ψ[σ(t)], t = 0} with respect to its wave dynamics {i.e., m ψ[σ(t)], t > 0)}. In other words, a t = 0, where particle m started to create a narrow wavelet of ψ[σ(t)] that moves away from m as depicted in Figure 2.5, which is the particle's dynamics m ψ[σ(t≈0)]. As time moves on [i.e., t > 0)], its dynamics decreases [i.e., |ψ[σ(t≈0)|> |ψ[σ(t)]|] since ψ[σ(t)] decreases with time [i.e., entropy increases]. This means that correlation efficiency between the wave dynamics at t = 0 with respect to t > 0 mass m is anticipated to be highest as given by,

$$\Phi = |\psi\,[\sigma(t>0)\,]|/|\psi[\sigma(t=0)\,]| < 1$$

where ||denotes as an ensemble average or equivalently we have,

$$\Phi = \Delta E(\Delta t > 0)/\Delta E(\Delta t \approx 0) < 1$$

In which we see that further away the created wave $\psi[\sigma(t>0)]$ moves away from the particle, the weaker its relationship to the particle's

dynamics (i.e., correlation efficiency). In other words, the more difficult to determine its particle's dynamics. For simplicity I had assumed m is constant (i.e., does not decay with time), since the smaller a particle is supposed to decay faster in time within our universe.

$\psi[\sigma(t)]$ originated by different sources

Wave dynamics $\psi[\sigma(t)]$ can be physically originated by different sources, such as particle annihilation, quantum leap, or by illumination, and others. But without a within a temporal (t > 0) subspace to subport, it cannot be created. Since $\psi[\sigma(t)]$ is a dynamic physical quantity than cannot be captured or isolated within a tiny space as particle. This is the reason to equate a package of $\psi[\sigma(t)]$ as a particle like photon is a serious mistake in modern physics. Since $\psi[\sigma(t)]$ is a package of conservative energy [i.e., $\Delta E(\Delta t)$] decreases naturally with time like any isolated subspace within our universe. From which it is apparent, without the support of temporal (t > 0) aspect of our universe, we cannot produce a wave within an empty space. This is one of the reasons why de Broglie's wave dynamics is virtual as mathematics since his principle was developed within a virtual mathematical empty space of which his principle is as virtual as mathematic. And this precisely the reason why de Broglie particle-wave is a purely mathematical principle, unless a temporal constraint is imposed with the principle, de Broglie particle-wave mechanics is mathematical virtual mechanics like Newtonian, relativistic, quantum mechanics, and as well all the new and old sciences are virtual mathematical principles. Since all of them were built on "the same" mathematical empty space platform.

Conclusion

Since origin of all our sciences were built on a virtual empty space platform, for which all our sciences are time independent or timeless (t=0).

This is precisely why all principles and laws of our sciences modern or classical are deterministic (i.e., absolute predictable). In this presentation I have shown Louse de Broglie's particle-wave dynamics is one of the many examples in our modern physics, is a typical a mathematical principle instead as a physical realizable principle. Since de Broglie's wave mechanics is one of the pillars in mathematical physics, this must the reason that his wave dynamics was the center of this presentation.

Nevertheless, this presentation shows only a small part of a mathematical principle can hurt to our science, since we are living in a temporal (t > 0) space but not in an empty virtual mathematical space. The apparent reason is that all our sciences were developed on the same empty space platform, although it is mathematical virtual platform, but it is where our mathematical sciences originated.

Since the damages have been done, but my question is thar; what are we going to do with our sciences? Although we should not blame the originators, but their innocence cannot be totally shadow by their mistake. Nevertheless, it must all about us, are we responsible scientists or just a group of renowned talk show physicists?

CHAPTER 21

Interpretation of ψ(t)

Wave dynamics cannot be mathematically simulated.

Since our time space has one and only one consequence

Since we are living in an energy conservation dynamics time space (i.e., our universe), everything changes with time being a forwarded variable coexisting with our universe that cannot be changed. Let me assume there is a particle (i.e., object) of mass in motion (i.e., assumed up and down) in very high speed which creates a wave ψ[σ(t)] e.g., EM wave. This is depicted in Figure 21.1 where σ is a spatial coordinate system that changes with time. Since wave ψ[σ(t)] is a byproduct of a particle's dynamics (i.e., motion), there exists a correlation of ψ[σ(t)] with respect to particle's motion. From this we see that the further away the wave ψ[σ(t)] travels its correlation with the moving particle that created ψ[σ(t)] will be weaking, since ψ[σ(t)] changes with time (i.e., entropy increases). Strictly speaking, even with the width of every wavelet of ψ[σ(t)] will change somewhat.

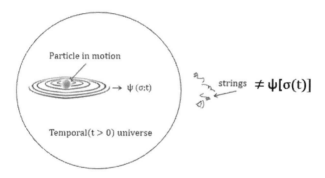

Figure 21.1 ψ[σ(t)] is a byproduct of particle's dynamics. It cannot be substituted by any mathematical equivalent. For example, String Theory

What m ψ[σ(t)] really means

Momentarily, let us assume m is constant [e.g., everything has a life span), for which its correlation between particle dynamics at t = 0 is given by {m ψ[σ(t)], t = 0} with respect to its wave dynamics to the particle's motion is given by {m ψ[σ(t)], t > 0)}. In other words, when t = 0, where particle m started to create a narrow wavelet of ψ[σ(t)] that moves away from m as depicted in figure 21.2.

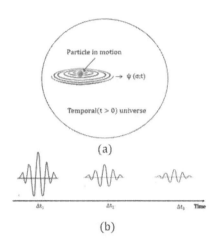

Figure 21.2: (a) shows a created wave dynamics of ψ[σ(t)] due to a particle in motion, (b) shows its strength changes with distance (i.e., time).

This represents the particle's dynamics m ψ[σ(t≈0)]. But as time moves on t > 0, its wave dynamics decrease [i.e., |ψ[σ(t >0)]| < |ψ[σ(t≈0)]| since ψ[σ(t)] decreases with time. This means that degree correlation between the wave dynamics at t > 0 with respect to t ≈ 0 particle's motion is anticipated to be lower as given by,

$$\Phi = |\psi[\sigma(t > 0)]|/|\psi[\sigma(t \approx 0)]| < 1$$

where || represents an ensemble average. In which we see that further away the created wave ψ[σ(t>0)] moves away from the particle (i.e., at t ≈0), the weaker its correlation to the particle's dynamics (i.e., correlation efficiency). In other words, it is more difficult for ψ[σ(t)] to determine the associated particle's dynamics (i.e., particle in motion). Again, for simplicity I have assumed m is a constant (i.e., does not decay with time), since the smaller a particle is it will decay faster with time.

Sufficient and Necessary Conditions for Wave creation

The necessary condition for particle dynamics to create a wave, is that particle must be situated within a subspace that can create wave. Which must be the sufficient condition. For which the subspace cannot be empty. Since the hypothesis is within our temporal space, it has time (i.e., t > 0) and substance (e.g., EM medium), which satisfies the sufficient condition to create ψ[σ(t)]. In which we see that a wave cannot be created within a virtually empty space.

Since wave dynamics is a byproduct created by a specific particle's dynamics, yet its association relationship decreases as wave ψ[σ(t)] propagates away from the moving particle. This means that that wave is unique which cannot be created by some other means. It is a mistake to assume that ψ[σ(t)] carries a full account of the particle in motion that is reference [1]. From this we see that it is just one of the countless types of evidence that our sciences new and old were developed on an empty space paradigm. But empty space is a virtual mathematical subspace that does not exist within our temporal universe.

[1] M. Born, *The statistical interpretation of quantum mechanics*, Nobel Lecture, December 11, 1954.

Particle-wave Dynamics Relationship

Since every wave propagates dependently with time, for which a correlation between them can be defined as given by,

$$\Phi = |\psi[\sigma(t > 0)]|/|\psi[\sigma(t = 0)]| < 1$$

The | | represents an ensemble average. This relationship can be expressed as,

$$\Phi = \Delta E(t > 0)/\Delta E(t \approx 0) < 1$$

In which we see the energy of ψ(t) degraded with time. This is precisely where coherence theory of light was derived, except it was developed from an empty space platform. Nevertheless, empty space has no time and no space. This is the virtual mathematical space from which all our sciences were developed. This is precisely why all our sciences are deterministic. Yet, science is not supposed to be precise.

Coherence Theory of Wave ψ[σ(t)]

To view the creation of particle-wave dynamics, let ψ[σ(t)] be divided equally into two waves $\psi_1[\sigma(t)]$ and $\psi_2[\sigma(t)]$ respectively as depicted in Figure 21.3. Since ψ2[σ(t)] travels a longer time (i.e., distance) to meet $\psi_1[\sigma(t)]$ at the observation screen p, we anticipate that $\psi^2[\sigma(t)]$ will degrade more than $\psi_1[\sigma(t)]$ because its entropy increases relatively higher. It is the visibility measure that is used to determine the coherence between these two beams of light. But within our time-space it is time that changes the property of light, not the speed of light that changes the time (i.e., time invariant).

Despite this, all our sciences, including both new and old, were developed on the same empty space platform, which is precisely the reason all our sciences are deterministic. In other words, they are timeless or time invariant sciences but not physically realizable. The reason why all our sciences seem to work in practice, is that we have bent the rule to make them approximated (i.e., avoided the law of time). The irony is that modern science wants us to make them exact and precise, yet it is our universe that
does not allow sciences to be exact and deterministic.

….Nonetheless, to make our sciences usable in practice we must view them as approximated. But modern sciences tell us to apply them exactly as the principles presented to us. My question is which one do you want to follow?

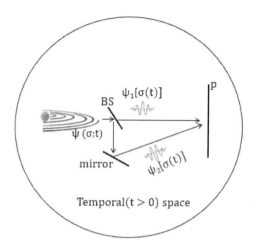

Figure 21.3 shows a beam of ψ[σ(t)] is divided ψ1[σ(t)] and ψ2[σ(t)] by beam splitter BS. P is an observation screen.

Back to Classical Coherence theory

Since all our mathematical sciences were built on a timeless empty sunspace platform, strictly speaking they are not physically realizable sciences. For example, classical optics or modern optics alike, they are time independent deterministic sciences (i.e., time invariant) like all other sciences. This is precisely why coherence between two divided beams are time invariant from the splitter instead of changing with time.
But within our time-space waves change with time with a coherence length between $\psi_1[\sigma(t)]$ and $\psi_2[\sigma(t)]$ instead of being determined by the changes with time. But $\psi_1[\sigma(t)]$ and $\psi_2[\sigma(t)]$ do not degrade with distances or sections of time (i.e., Δt_1 and Δt_2) of $\psi_1[\sigma(t)]$ and $\psi_2[\sigma(t)]$ that it had traveled. This Changes contradicts the classical coherence theory, which

is time invariant in space with its subspace model is empty space.

Ψ(t): The Legacy of Hamiltonian Mechanics

Albeit the first implementation of ψ(t) may be traced back to William R. Hamilton's mechanics. However, it must be due to de Broglie particle-wave dynamics that made it more prominent. Like Dirac, Schrödinger, and others to follow, all of them were mathematically implemented wave dynamics of Ψ(t) which has nothing to do with the physical of our universe. Since all our analytical sciences were all built on the top of a virtual empty space platform, all their wave functions are either deterministic or mathematically virtual. This means that all the legacies of ψ(t) after de Broglie have been wrongly interpreted as they are purely virtual mathematical interpretations. But they have suffered severely from a physical realizability standpoint. For decades or over a century, we have had accepted that ψ(t) is a physically realizable predictor, that can precisely predict the particle's behavior where ψ(t) is associated with either actively dynamic particles or by passive illumination. But in practice, all physically realizable principles must be approximated.

Ψ[σ(t)] Is an Energy Conserved Dynamic Time-space

Like the creation of our dynamic universe, wave dynamics Ψ[σ(t)] is an energy conserved dynamic stochastic time-space as depicted in a two-dimensional format shown by Figure 21.4. From which a group of limited ripples

$$\{i.e., \psi^2[\sigma(t)] and \, \psi^2[\sigma(t)]\}$$

travel outward with time. But their energy profiles decrease with it as they expand but are also conserved. This is like the dynamic expansion of our

universe, with ripples of packages separated by conserved energies.
….Once again, we have seen that wave $\Psi[\sigma(t)]$ is an energy conserved dynamic time-space that behaves somewhat like the dynamic expansion of our universe. We see that $\Psi[\sigma(t)]$ is a conserved dynamic stochastic time-space that changes with time. Note that it is not a static deterministic process as developed from an empty space paradigm, as all sciences have done. Albeit once again, $\Psi[\sigma(t)]$ is a byproduct of a physical particle dynamics which cannot be substituted by a set mathematically simulated equivalence such as String Theory [2].

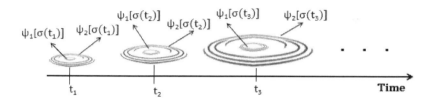

Figure 21.4 shows a set of wavelet dynamics ψ1[σ(t)]and ψ2[σ(t)]} expanding at speed of light with time. In which their profiles degrade with time, but their energy is conserved.

It has no physical connection to justify that the theory exists within our time space. That is difficult to reconcile that mathematically equivalent theory exists a correlation (e.g., genetic trace) with a particle's dynamics that created the wave since $\Psi[\sigma(t)]$. By the way $\Psi[\sigma(t)]$ is a byproduct of a particle's dynamics which cannot be simply created. For which String Theory is not a physical realizable science, but a mathematical gimmick. Albeit why do we keep promoting it?

[2] B.. Zwiebach, *A First Course in String Theory* 2nd ed. Cambridge U. Press(2009).

Introduction to Physically Realizable Physics

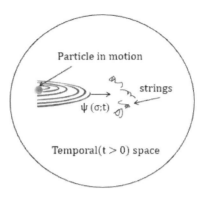

Figure 21.5 shows a set mathematically equivalent springs simulate $\Psi[\sigma(t)]$, but it does not exist a correlation with its particle's dynamics.

$\Psi(t)$ Can Have Different Forms

Nevertheless, particle-wave dynamics $\Psi(t)$ can be induced by different forms of energy. Example such as particle annihilation, EM radiation, by illumination, and others. Since $\Psi(t)$ is a physically quantity that cannot be simply substituted by mathematics or computer simulated substance that does not have any natural association (i.e., correlation) with the dynamics that created the wave $\Psi(t)$. This is a unique feature since our universe has one and only once physical reality. Once it has gone by it will never be repeated. And this is the subspace we live in which cannot be substituted by mathematics or by fantasy computer simulation. Although science needs mathematics to facilitate its complicity description, but mathematics is not science.

Nonetheless, let me show how $\Psi(t)$ can be created by means of particle annihilation as depicted in Figure 21.6. Which is basically the same model as our universe was created. Notice that temporal ($t > 0$) universe is closer to the truth model, since it does not violate any known law of nature (e.g., law of time and others).

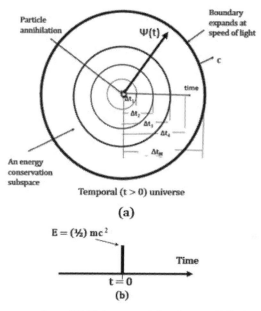

Figure 21. 6 shows a creation of Ψ(t) by a particle of m annihilation. Where m is the mass.

Created by a Sequence of Impulses of Energy

Similarly, as shown in Figure 21.7, a sequence of impulses energy creates a sequence of wave dynamics $\psi_1[\sigma(t)]$, $\psi_2[\sigma(t)]$, ... respectively. Once again, all the wave dynamics {i.e., $\psi_1[\sigma(t)]$, $\psi_2[\sigma(t)]$, ... } are the byproducts of the corresponding energy pulses. They exist a one-to-one correspondent correlation which cannot be altered. In Without hypothesizing its specific particle-wave dynamic, it is impossible to distinguish all the correspondent wavelets {i.e., $\psi_1[\sigma(t)]$, $\psi_2[\sigma(t)]$, ... }.

They will be instantly (i.e., $t = 0$) and simultaneously superimposing, as the superposition principle of Schrödinger tells us. This is one of many examples to show why our modern physics is so wrong. It is currently out of control since millions of mathematical physicists young and old alike keep promoting it. This never ended vicious cycles are more harmful than beneficial.

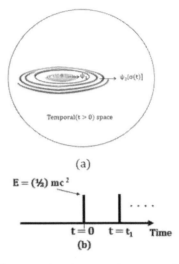

(a)

(b)

Figure 21.7 shows a set of nonoverlapping wavelets as produced by a sequence of well separated particles' annihilation.

Ψ(t) Cannot Be Mathematically Created

Laws of nature are for us to discover but not to create. The correlation between two eigen wave dynamics is a part of nature, which cannot be mathematically substituted. For example, your children and grandchildren have genetic traces of yourself, they cannot be mathematically simulated or even adopted. String Theory [2] is one of the many examples, that is mathematically correct, but is not physically realizable. For example, wave dynamics of Ψ(t) cannot be mathematical simulated, since Ψ(t) is a byproduct of a particle's dynamics. It exists a highly genetic code (i.e., correlation) relationship with the particle's dynamic that created the wave, which cannot be substituted. This is related to our temporal (t > 0) universe, which has only one physical reality that cannot be repeated [3].

[2] B. Zwiebach, *A First Course in String Theory* 2nd ed. Cambridge U. Press (2009).

[3] F.T.S. Yu, *"Origin of Temporal (t > 0) Universe"* CRC Press, Boca Raton, Fl (2020).

Conclusion

I have shown that a wave $\psi[\sigma(t)]$ is a byproduct of a particle's dynamics (i.e., motion), and exists a unique correlation of $\psi[\sigma(t)]$ with the particle's dynamics that created the wave. For this wave $\psi[\sigma(t)]$ cannot be mathematically equivalenced to or computer generated. It is a byproduct from a particle's dynamics only. Since wave is a part within our time-space which has one and only one occurrence that cannot be repeated. For example, an adopted child has its own genetic code, which cannot be mathematically equivalence to or computer simulated from her or his adopted parents.

Finally let me stress that, damages have been done to our modern science. It is my question as to how far we will let it continues? Since it is not where it originated to be blamed, because it was also where our science originated. But one question remains: we responsible scientists or a group of supporters for all those renowned talk show scientists? Believe it or not, turn on any science channel to find out.

Chapter 22

Why ancient Chinese did not have a fundamental science?

 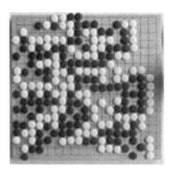

But ancient Chinese did invented Abacus, Go, and others.

Why ancient Chinese did not have a fundamental science?

Two of the most important human sensors must be our eyes and ears, where Chinese characters were developed based on light transmission, while western cultures were developed from sound transmitted language. This must be the reason why western culture is more inclined to develop alphabetical type languages. While eastern civilization (i.e., Chinese) developed a pattern form written language. Yet, mathematics is a symbolic substituted language, for which we see that it is more inclined for westerners to develop mathematics than for easterners (i.e., China). For example, as can be seen in Figure 1 by a set of Chinese characters with respect to its translated English counterpart.

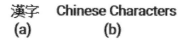

漢字　　Chinese Characters
(a)　　　(b)

Figure 1(a) shows a set of Chinese written characters with its English translated counterparts. (a) Chinese characters, (b) The corresponding English translation.

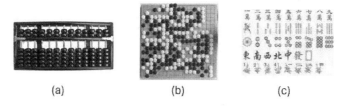

(a)　　　　(b)　　　　(c)

Figure 2 shows some of the ancient Chinese invented devices, (a) abacus, (b) Go, and (c) Mahjong.

From which we see that although the Chinese has an uninterrupted civilization over five thousand years, yet she has failed to create the kind of mathematics that western civilization has developed. Since fundamental science (i.e., theoretical physics) is mathematically oriented, this is precisely the reason why ancient Chinese virtually did not have an analytical (i.e., mathematical) science.

Since written language is an important attribute for a civilized culture, without a sophisticated written language it hardly can develop into a higher civilized society. Western European languages were developed primarily based on syllabic substituted language (i.e., sound), while Chinese written characters were created based on pictorial format (i.e., light transmission) which is a single syllable tonal language. This must be the reason that gave Europeans a more natural inclination to create a symbolic substituted mathematics. Apparently, it is very unlikely to implement Chinese characters for a symbolic substituted mathematics. Nevertheless, science is mathematically symbolic presentation, which is precisely the reason why the ancient Chinese virtually did not have science.

Symbolic representation of science

Without the implementation of symbolic substitution, principles of science would be very difficult to facilitate. For example, how can one communicate chemical elements by Chinese characters. In which we see that without using symbolic substituted chemical elements (e.g., O, H, C, …) it will be more difficult to communicate with Chinese language.

Moreover, without those simple mathematical attributes (i.e., $\int$, $\sum$, Δ, $\approx$, …), it will need a lot of more description to explain in plain texts. In which we see that it will take a lengthy set of Chinese characters to facilitate simple laws of science as depicted in Figure 3.

$E = mc^2$ 能量等於重量乘以光速的平方

(a) (b)

Figure 3 shows a simple symbolic equation [i.e., (a)], but it will take a lengthy set of Chinese characters to describe, as depicted in figure (b).

Since fundamental laws and principles are presented in symbolic substituted form, it can be seen why it is extremely difficult to describe them in Chinese written language. This is precisely why ancient China virtually did not have mathematic, because it was very unlikely and unnatural for Chinese to adopt a set of mathematical attributes to create mathematics,

Ancient Chinese knew addition, subtraction, multiplication, and division, and had invented the abacus to facilitate the calculations. Yet Chinese failed to symbolize +, -, x, and ÷. Even though ancient Chinese did have arithmetic, yet it is not even close to the number theory of the west.

Nevertheless, ancient Chinese had invented several viable technologies such as the compass, paper, dynamites, and rockets, Chinese had failed to discover principles of fundamental sciences (e.g., gravitational law, law of entropy, law of energy, and many others). The failure to discover fundamental principles were mainly due to Chinese written characters which is not a type of alphabet like European languages.

Nonetheless, if Chinese had created a set symbolic attribute needed for mathematics, then there has no reason why Chinese would not invent the kind of science like westerners. From which we see that it is her pictorial characters that had prevented ancient Chinese from developing the kind of symbolic attributes for her mathematics. Yet, this is the kind of written language that holds China together as a nation and as a people. Without it, China may not have held together as a nation and as a unique civilization for thousands of years. It is more likely that it would fragment into several nations of different dialects as European nations.

Nevertheless, science is a language speaking for the truth. Any disadvantage is not meant to be inferior. In contrast, it makes a civilized nation smarter.

Conclusion

In conclusion we see that it is the attributes of Chinese written languages had prevented ancient Chinese from inventing mathematics. But it is by no means a disadvantage in terms of mathematical science. On the contrary, China is one and only one human civilization that has survived as a nation and as a people for over five thousand years of uninterrupted history. By transforming Chinese culture and importing mathematical sciences from the west, China has benefited its people and as well as a nation. From which we see that a great advance has been taking place in modern China without losing a major part of its Chinese cultural transformation. Surprisingly, it turns out to be a better China for her people and for the world, by not giving up her fundamental cultural heritage for the sake of importing mathematical science.

Chapter 23

AI is at different level of knowledge abstraction.

AI needs Associative, cognitive, innovative, and creative knowledge base abstraction.

AI is at different level of knowledge abstraction.

Artificial intelligent (AI) must be developed by matured technologies, but it should not be based by fantasized principles of science. In other words, AI cannot be a mathematically simulated science such like quantum communication or computing is a virtual nonexistent promise. Unlike fundamental science, AI must be a physically realizable mechanics. In other words, AI is a technological science but not a fundamental physics. For an AI to be realized, there is a series knowledge-based abstractions needed to overcome. For examples associative, cognitive, innovative, and creative knowledge abstraction. For which the success of AI is depending on each level of knowledge base to pursue. For example, as from lower associative, to cognitive, innovative, and to creative knowledge base abstraction (i.e., a human like). From which we see that, each level of knowledge becomes more complex and sophisticated. Since without this differentiation, the search on AI will lead to simulated fantasy, which is not the type of AI that real scientists are looking for.

Human base knowledge abstractions

Since AI is supposed to be a realizable mechanics behaves like a human. So, what are the human base knowledge abstractions? As from my understanding we human has at least four important knowledge base intelligences, associative, cognitive, innovative, and creative. And these levels of complexity increase exponentially from associative to creative abstraction. From which we see that for an AI to be successfully realize it will take a tremendous effort to accomplish. For example, just base on associate knowledge AI is easier to realize than by cognitive. From which we that it is anticipated to be more difficult to develop on innovative and to creative human like AI.

What is an associative knowledge?

Associative memory knowledge is based on associating with something that easier to remember. For examples, to remembering a Chinese or Japanese name are usually based on meaning rather than how to pronounce a name. To remembering a telephone number is based on sequencing, symmetric property, partition, and other to remember.

For instance, I used to remember Professor Nagakawa Kazuo first time we met. Since I neither spoke nor read Japanese that I had a hard time to remember his name on various occasions we met again. It was kind of embarrassing to constantly ask his name, until I read his name card in Japanese that is like Chinese. His name is equivalently means that a gentleman is swimming in a river. From that I still remember his name today who I have not seen for over 30 years.

What is a cognitive knowledge?

Another level of knowledge base is the cognitive abstraction or the art of understanding. For examples; understanding the axiom of a multiplication instead of memorizing a mortification table. Similarly understanding the physical significant of Einstein's energy equation (i.e., $E = mc^2$) and others. Basically, cognitive knowledge abstraction is understanding how it work but not just remembering the equation. For example, how many physicists know the significant of entropy and how entropy must increase with time. For instance, how many physicists know entropy is an energy degradation principle. The reason for asking these questions it that, otherwise theoretical physicists would not had have developed modern physics (i.e., wrong!).

What is an innovative knowledge?

Another level of knowledge must be the innovative abstraction which is understanding why thing works. For example, take Einstein energy equation, it is the existent of time in space that annihilates a mass to energy (i.e., $E = mc^2$), otherwise mass m has no way to annihilate. Another example is why entropy must increase with time, regardless we like it or not. It is because our universe is an energy conservation dynamic expanding subspace, for which its energy degrades with time.

What is a creative knowledge?

Which is an ultimate human like knowledge that we would like to emulate for an AI. Aside the required associative, cognitive, and innovative knowledges, human can create new knowledge that most of the animals on the planet earth still unable to match. This must be the highest level of knowledge base (i.e., creativity) that a human can achieve, since human is the only animal (i.e., biological AI) has this kind of ability.

Intelligent knowledge base abstraction

Figure 1 shows a hierarchy knowledge base diagram, in which we see that creative knowledge is the highest abstraction that human can achieve.

Figure 1 shows different level of knowledge base abstraction.

Necessary requirement for AI to success?

In view of knowledge base requirement for AI to success is dependent on the success of various levels of knowledge base abstraction; associative, cognitive, innovative, and creative. From which every AI is not meant to achieve the highest creative abstraction, but it is limited by current available technological gadgets. In other words, it is wrong to promote an AI beyond the current technological limit such as fantasied quantum computing and particle travels at speed of light. Nevertheless, AI is supposed a physically realizable mechanics that we human would like to create, but not a fantasy AI that beyond the current technological limit that can achieve.

Conclusion

In conclusion I have shown knowledge is based at different level of abstraction; associative, cognitive, innovative, and creative. AI is supposed to be a physically created mechanics (i.e., being) but must obey the laws of nature, that cannot be fictitious as mathematical sciences is.

For an AI to be successfully developed is very much depending on the availability of current technological gadgets, but not to include the fantasy laws of sciences that have not shown physically realizable evidence. For examples, such as substances travel beyond the speed of light, instantaneity, simultaneity interconnection, changing time, quantum computing, and others. Finally, how many times Einstein, Schrödinger or you had cheated you?

Chapter 24

Theory of temporal (t > 0) Space

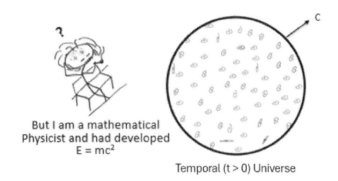

Temporal (t > 0) Universe

How many theoretical physicists knew mathematical physics is **not** a physically realizable physics?

Theory of temporal (t > 0) Space

Mathematics is not science, but science uses mathematical language to speak the truth. It is reasonable to develop a temporal (t > 0) space theory like theory of set of mathematics. For which let me begin within a temporal (t > 0) space, as depicted in Figure 1 in which it is filled with substances that coexist with time. This new concept of time-space theory is appropriated for application to our physically realizable sciences. This temporal (t > 0) space (i.e., our universe) is closer to the truth instead of using Einstein's 4-d spacetime continuum that violates all the laws of nature.

Nevertheless, theory of set was developed strictly from a pure mathematic standpoint, but mathematics is not science. For which Temporal (t > 0) space theory is a useful time-space for realizable mathematical science. It is because mathematical physics is not a realizable science, since mathematics does not need to obey laws of nature.

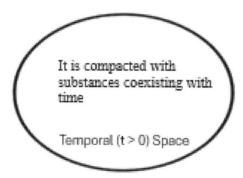

Figure 1 shows a temporal (t > 0) bounded subspace.

Theory of set

Theory of set was developed from an abstract empty space paradigm as all our mathematics did since mathematics was created. But set theory is a pure mathematics that fundamental physicists should not have used.

For example, empty set is not a physically realizable set as depicted in Figure 2. In which we see that an empty set (i.e., equivalently to an empty subspace) is an acceptable set within a universal set. But it is not physically realizable set for science since it is not a subset within a temporal (t > 0) set. Theory of set is pure mathematics which does not need to obey laws of nature. For example, the famous Russell's paradox as I quote, "It does not exist a set of all set", in which we see that mathematics does allow paradox, yet science does not.

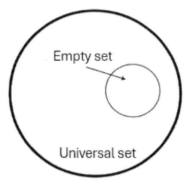

Figure 2 shows an empty set can be situated within a universal set since mathematics does not need to obey laws of nature.

Where did all our sciences develop?

Science uses a mathematical speaking language and need to obey laws of nature which cannot be fictitious as mathematics is. For example, as depicted in Figure 3 believe it or not, origin of our mathematical science was developed from an empty space substrate. Which is precisely why our analytical sciences are fictitious as mathematics. Strictly speaking our mathematical physics is not physically realizable. In fact, all mathematically oriented science is mathematics but not science. Science does not allow paradox to exist, but mathematics does. For example, the paradox of Schrödinger's cat has been proven false since science does not allow paradox. From which we see that empty space paradigm is not a physically realizable platform that can be used in science. Yet mathematical physicists have used it since the dawn of our analytical science.

Figure 3 shows an empty space which has no substance in it and no boundary. It is a mathematical subspace.

Theory of a physically realizable space

Since our time-space is a 3-dimensional space coexisting with time, where time is a forwarded variable [i.e., temporal (t > 0)], conventional set theory cannot be used to emulate our temporal (t > 0) universe. Everything within our universe changes with time. This slide shows more evidence that conventional mathematics which includes set theory cannot be simply employed in science.

Although, science uses a mathematical descriptive language as depicted in Figure 4, we see that an empty set is equivalent to an empty subspace. This cannot exist within a temporal (t > 0) space known as temporal (t > 0) exclusive principle [1].

[1]. F.T.S. Yu, "Time: The Enigma of Space", Asian Journal of Physics, Vol. 26, No.3, 143-158, 2017.

Introduction to Physically Realizable Physics

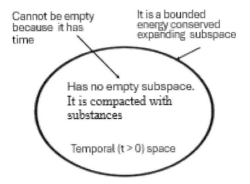

Figure 4 shows a physically realizable bounded space, in which everything within the space changes with time.

From temporal (t > 0) space to creation

One of the best astrophysics observatories must be the thousands of years accumulated observation by us. Any created universe by mathematical oriented physicists is very unlikely to be truth since mathematics is not science. In which one of the worst models must be the 4-d zero-summed spacetime continuum of Einstein. Unfortunately, space scientists have used it for over a century. Surprisingly, they did not even know that this model violates all the known laws of nature.

Nevertheless, one may ask, even though Einstein's relativity theory is false, why is his energy equation correct? Ironically, $E = mc^2$ was derived from his false relativity theory. The answer is that every mathematical science has multiple solutions whether they are correct or not. Yet it is the physical realizability constraint [i.e., law of temporal (t > 0)] that limits the numbers of possible solutions to one correct answer. Only one physical reality can emerge from a physical realizable mathematics at a given time.

Because of the limited knowledge of our universe, one of the best models of it must be the creation from a mass to energy transformation [i.e., $E = (½) mc^2$]. This creation was started within a space that has time

[I.e., temporal (t > 0)], otherwise it would be more difficult to justify how a mass can be annihilated by itself without time. In view of an accepted kinetic energy equation [$E = (½) m v^2$], we see that its total energy from a given mass is equivalent to $E = (½) m c^2$ where m was annihilated means totally changed into EM wave at speed of light. Notice that this equation is virtually identical to Einstein's energy equation, except his energy equation was derived from his relativity theory which is not a physically realizable theory [2]. This is also one of the types of evidence shows every mathematical problem has muti solutions. Nevertheless, a physically realizable solution has only one solution.

The supposed creation is the limitation of our present knowledge of creation. But this temporal (t > 0) universe model obeys all the laws of nature that I have known. For example, we have seen Einstein's 4-d spacetime has already created scores of fantasy principles that have been polluting our modern science sector. Why then does the scientific community continues to believe his 4-d spacetime continuum is truthful?

[2] F.T.S. Yu, "Introduction to Physically Realizable Physics" ISBN 9798877098312, chapter 13, (2024).

Why every mathematical problem has multiple solutions?

Every physically realizable science must result in one solution, even though every mathematical physics has multiple solutions. Since mathematical physics does not need to obey laws of nature, this is precisely why mathematical physics is not science. Besides, physically realizable science does not allow paradox while mathematics does. But science used a mathematical language speaking for the truth, yet the truth must be physically realizable. For examples, paradox of Schrödinger's half-life cat, time dilation of Einstein's relativity, Quantum electrodynamic of Feynman, and many others are not physically realizable sciences. The apparent reason is that our universe takes everything within her at expanding her boundary at speed of light. For which we see that why her entropy increases (i.e., enclosed energy degrades) with time as depicted in Figure 5 in next slide.

In other words, everything within our universe changes with time (i.e., its energy degrades) as our universe expands at speed of light. In which we see that our universe is an energy conservation dynamic subspace. Mathematical physics is not a physically realizable science since it simultaneously produces multi solutions. But physically realizable mathematical science can only have one solution.

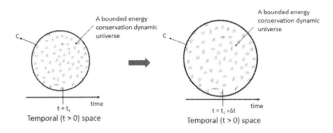

Our temporal (t > 0) universe

Figure 5 shows our universe exists at time $t = t_1$ and it enlarges a $t = t_2$. Yet the total energy within our universe is conserved.

Disprove Einstein's relativity.

Since Einstein's relativity theory is strictly mathematical, for which we anticipate that it has multi solutions. For example, a moving clock is travelling from San Diego to San Francisco at a constant speed of v as depicted in Figure 6. According to Einstein's relativity, time clicks slower for a moving clock than the stationary one. From a mathematical standpoint, it has a great number of solutions for the moving clock to reach SF travels at various velocities of v. Since relativity theory is mathematics, it has a huge number of solutions. One of the solutions is that instantly (i.e., at the expense of time) the moving clock will reach SF if the moving clock is moving at speed of light (i.e., $v = c$).

Figure 6 shows a moving clock travel at a velocity of v from San Diego to San Francisco.

However, since physically realizable science has only one solution, if the same scenario is submerged within a temporal (t > 0) space platform as depicted in Figure 7, it has only one solution. In which we have found that the moving clock ticks at the same rate as the stationary clock, even if the speed of the moving clock reaches the speed of light. This proves once again that Einstein's theory of relative is false since we cannot change the time of our universe.

Temporal (t > 0) space

Figure 7 shows the same scenario but submerges within a temporal (t > 0) space. From which we see that both clocks tick at the same rate.

Einstein relativity was mathematically created!

Since Einstein's relativity theory was derived on an empty space paradigm like all our fundamental laws of physics, his relativity theory does not need to obey the laws of nature [i.e., temporal (t > 0)]. In other words, Einstein's relativity theory (i.e., special, and general theory) is mathematical create. Nevertheless, his relativity theory is a mathematical problem, which is anticipated to have several solutions regardless of whether they are physically realizable or not. That is precisely why his energy equation turns out to be correct, in addition to a bunch of other fictitious and false solutions (e.g., rest and move masses, and others) they were derived from his relativity theory. Yet, without the law of temporal ($t > 0$), mass has no way to annihilate. Since relativity theory is mathematics, it has a number of solutions but not all of them are correct solutions. For example, one of his energy $E = mc^2$ is correct and all the others are not. In which I have shown a correct solution can also be produced from incorrect principle such as his relativity theory, since mathematical science is still a virtual mathematics. However, if a constraint of physically realizable condition is imposed on mathematical science, then a physically realizable mathematical physic will be is a realizable physics. In other words, as I have shown if a hypothesis developed on a temporal ($t > 0$) platform, then its mathematical solution will guarantee be physically realizable within our time-space.

Why mathematical physics is not science?

Since mathematics was man made, it does not need to obey the laws of nature such as temporal ($t > 0$). Although science uses a mathematical language, but it must obey laws of nature. Laws of nature are supposed to discover but not to create. For examples, Einstein's spacetime and as well Schrödinger 's quantum theory was mathematically created. They are mathematical sciences, which are not physically realizable. Which is exactly why all our mathematical physics are not physically realizable. One simple example is that mathematical physics allow. For paradoxes (e.g., time travelling, superposition, and others) while physically realizable physics does not.

Since physically realizable physics has one physical reality, this is precisely why theory of temporal (t > 0) physics is crucial. Otherwise, we would continuingly be buried by mathematically oriented sciences. If we do not do anything about it, the damage will be more server. And it is not fair to our students, children, grandchildren, our young physicists, and to our conscience.

A universe of all universes does not exist.

Within the theory of temporal (t > 0) space, we see that a universe of all universes does not exist, since physically realizable universe has only one universe which cannot be repeated. This theorem is like Bertrand Russell's paradox in set theory, but the implication is rather different. Since set theory is mathematics, it follows the mathematical logic. It has nothing to do with time, energy, entropy, and others. But within our temporal (t > 0) space, everything changes with time. In other words, things within our universe must obey all the known laws of nature [i.e., law of temporal (t > 0), law of entropy, and others]. In which we see that a universe to include our universe cannot exist. Since this is the only universe that we have and cannot be repeated, it is a mistake to advocate paralleled or multiple universes.

A physically realizable paradigm

Science uses a mathematical language speaking for the truth (i.e., physically realizable), but mathematics does not need to follow the laws of nature [e.g., law of temporal (t > 0)]. For which mathematical science needs a realizable paradigm to guarantee that her solution will be physically realizable. Mathematical set theory (i.e., theory of set or space) is not a suitable theory that can be applied in our temporal (t > 0) space (i.e., our universe). By which I have shown a temporal (t > 0) space theory which follows the laws of nature. If mathematically oriented physicists use this theory as applied to his/her mathematical physics, its solution would be guaranteed to have only one temporal (t > 0) solution that obeys the laws of nature. Since within out temporal (t > 0) universe it has only one physical reality which cannot be repeated.

Conclusion

Temporal (t > 0) space theory is a useful time-space of set for realizable mathematical science. In conclusion I have shown that a temporal (t > 0) space theory like theory of set of mathematics which can be developed. Nevertheless, theory of set was developed strictly from pure mathematics standpoint, but mathematics is not science. For which it is because mathematical physics is not a realizable science. since mathematics does not need to obey laws of nature. Since every mathematical science has a few solutions regardless there are correct or not. But physically realizable mathematical science is limited by one. For which we see that physically realizable mathematics is a correct mathematical physics to explore, by which it offers only one physically realizable solution. Although temporal (t > 0) theory is like theory of set, I have shown that, within our time-space it does not exist a universe of all universes, since our universe has only one temporal (t > 0) universe, that cannot be repeated.

Chapter 25

Dirac's antimatter hypothesis is not science.

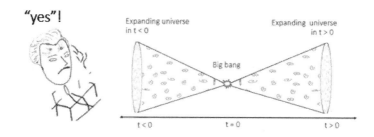

Firstly, are we living in a zero-summed universe? Dirac said "yes"! But it is a mathematical gimmick?

One of the best observatories must be everyday consequences that we had encountered in the past thousands of years history. We knew physically we change with time, but not changing time. From which we knew that every physical aspect attached with us goes with us, it is impossible for us to leave one of our arms behind, even severely amputated. From which we see that the amputated arm may not attached with us but follows us to the next day and after.

For year we wander about ourselves physically associated with time, we knew today is a different day from yesterday and tomorrow will be another day of uncertainty. Although constantly wondering about our space and time for thousands of years, why we have opted to accept a mathematically oriented physicist told us our past observations are wrong. It seems to me we have treated mathematics as science, but mathematics does not need to obey the laws of nature.

A zero-summed universe model

Figure 1 depicts a well-accepted zero-summed universe model that it violates all the laws of nature. This is one of worst model that have been used for overs a century. This means that all the past substances remained within our symmetry universe and waiting for us to visit?

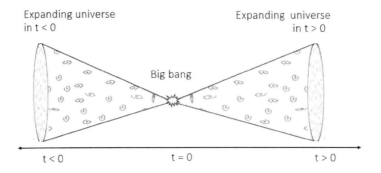

Figure 1 shows a zero-summed universe model where Dirac equation was developed from. This is essentially the Einstein's 4-d spacetime continuum that had been used by many theoretical physicists for over a century.

Dirac a mathematical oriented physicist

In view of Dirac's analysis, he is more a mathematical physicist than a physically realizable scientist. His derivation was so entrenched with mathematics than physical attributes. His apparent error can be depicted in Figure 2(a) where speeds of all the particles were missing. Secondly, if we insert a universal clock within his hypothesis as depicted in Figure 2(b), it is simply cannot exist within our temporal (t > 0) space since substance and emptiness are mutually exclusive.

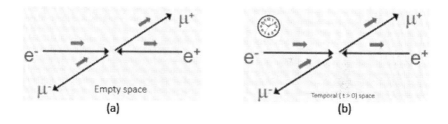

Figure 2(a) shows Dirac's hypothesis was developed within an empty space paradigm. which is nonphysical realizable paradigm since empty space and substance are mutually exclusive [i.e., temporal (t > 0) exclusive principle] as shown in Fig. 2(b).

Secondly, how can particles move within a space which has no time since it is time moves the particles. Dirac's equation was derived from an empty space paradigm, his equation is time independent that violates temporal (t > 0) law of space. For example, within our temporal t > 0 space, entropy increases with time, chances for particles to meet is uncertain particularly for smaller particles. Furthermore, without the existent of time [(i.e., temporal (t > 0)] particles cannot move.

A correct paradigm for Dirac

On the other hand, if Dirac's hypothesis is submerging within a temporal (t > 0) space paradigm as depicted in Figure 1(b) in previous slide, where e⁻ and e⁺ are the negative and positive charged particle, μ^+ and μ^- are the muons. With reference to a temporal (t > 0) space model as presented, I would challenge mathematically oriented physicists to evaluate Dirac's postulation if they could.

Nevertheless, with the insertion of a universal clock, under the law of temporal (t > 0) constraint, please show me Dirac can bring particle e⁻ and e⁺ together? Furthermore, the assumed μ^- and μ^+ must also obey law of temporal (t > 0). In other words, if his equation is correct, consequence from his postulation can only happen after the casualty [i.e., temporal (t > 0)], that cannot happen at current moment of time t = 0 or even ahead of time (i.e., t < 0). This is the primary reason that I do not have to disprove his equation mathematically since his hypothesis had firstly violated the law of temporal (t > 0). Overall, without a physically realizable paradigm, is very unlikely to obtain a physically realizable solution since science is a mathematical language but speaking for the truth.

What Dirac equation means?

It is not a surprise that Dirac equation is a zero-summed, non-energy conservative equation as given by,

$(i\partial - m)\psi(t) = 0$

Since his hypothesis was built on a zero-summed empty space, this is precisely why Dirac had postulated antimatters that exist in the negative time domain, as in contrast with our temporal (t > 0) universe, only exist at current moment of time (i.e., t = 0).

Nevertheless, if Dirac equation is written by a non-zero summed form as given by,

$i\partial\psi(t) = m\psi(t)$

I am sure he might not have postulated that anti-matters exist in the negative time domain, since there has no physical space (i.e., null) behind or ahead of current moment of time (i.e., t = 0). Even by his complicated analyses, mathematics alone cannot alleviate a non-physically realizable solution.

What is an energy conservation universe?

Firstly, let me show our temporal (t > 0) universe creation as depicted in Figure 2(a), we see that any substance within our universe goes with the universe (i.e., time). Which means that everything within our universe degrades somewhat with time. In words, it is the amount of degraded energy (i.e., work done) that fuels the life of our universe which includes us. Since entropy within our universe increases with the expanding universe, as depicted in figure 2(b), our universe is an energy conservation dynamics subspace.

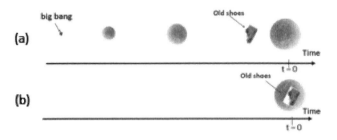

Figure 2(a) shows that everything goes with the universe, even an old shoe cannot leave behind as depicted in (b) since there is no space before or after the current moment t = 0 universe. Our universe has one and only one physical consequence at current moment t = 0.

What a temporal (t > 0) universe is?

Figure 3 depicted a composited diagram of our temporal (t > 0) universe which is situated within a larger temporal (t > 0) space which the closest to the truth model that I have known since it does not violate the known laws of nature (i.e., law of time, law od energy conservation, and others). In which we see that our universe was created by a huge energy [i.e., $E = (½) mc^2$], which is like an enlarging hot air ballon. From which we see that, Sooner or later the expanding universe will exhaust all its energy that created her by continuingly energy degrading (i.e., entropy increasing) with time. That is the end of our universe as we see that all her degraded energy will be deposited within the larger temporal (t > 0) space that our universe once embedded in. In which I have shown this is one of **the best** universe creation models that obeys all the known laws of nature.

Introduction to Physically Realizable Physics

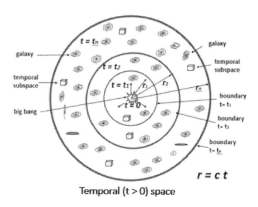

Temporal (t > 0) space

Figure 3 shows a time composited temporal (t > 0) universe. In which we see that our universe. was created within a preexistent temporal (t > 0) space. Otherwise, big bang initiation cannot be started.

What Dirac anti matter means?

Furthermore, let me take the Dirac equation as given by [4],

It is a zero-summed equation, which imply that anti-energy and anti-mass exist ahead of $t = 0$ (i.e., $t < 0$). But our universe is going with time. Nevertheless, those anti-matters or anti energy were supposed to have exited in the negative time of our universe that violates the law of energy conservation that created our universe. Then how can Dirac justify anti-matters within our current time-space? For which his equation is wrong even though his mathematics is correct. Mathematically speaking zero-summed and non-zero summed equation are the same, but the interpretation is different as from science standpoint. Nevertheless, if Dirac is written in a non-zero summed form as given by,

which it is very difficult for us to accept his anti-matter postulation. From which we see that how mathematical physicists had done to our science, since science is mathematical language supposed speaking for the truth.

[4] A. Hirschfeld, The Supersymmetric Dirac Equation, Imperial College Press, London (2011).

Conclusion

Since Dirac was one of the legendary mathematical oriented physicists, who had established a conner stone for mathematical physics, without the elimination of his work it will be difficult to move on for a physically realizable physics. It seems to me Dirac's mathematical physics and realizable physics are mutually exclusive. As from physically oriented standpoint, Dirac's is a bunch of mathematical analyses, it does not offer any physical realizable attributes to science but just mathematics. Which turns out more damage than beneficial. Nevertheless, Dirac's mathematical gimmick is even worse than Einstein's relativity theory, of which Einstein had produced a correct equation (e.g., $E = mc^2$) but Dirac's has none. Dirac's work is one of the many examples, that mathematical physics has confused the essence of physical reality of science. Like quantum theory of Schrödinger, it has more mathematics than physical reality. It seems to me that mathematically oriented physicists do not understand our own self is a subspace that goes with time but not changes time. For instant, once we have gone by time, but we still trapped within the dynamic universe. It is not possible to left behind time because there has not even an existent space for us to be left behind.

Furthermore, mathematical physics have done more damages than beneficial to our fundamental physics. Since science is a mathematics but speaking for the truth. Believe me or not, tune on any of those science channels to find out?

YouTube Links for narrations (by chapter)

CHAPTER 1: Discovery of Temporal (t > 0) Universe. Narrated by Lucy C. Yu

https://youtu.be/fLupEGpTPLY

CHAPTER 2: The Fate of Schrödinger's Cat. Narrated by Ann G. Yu

https://youtu.be/8Kiqy_rP-vE

CHAPTER 3: What is "Wrong" with Theoretical Physicists. Narrated by Ann G. Yu

https://youtu.be/Z_oI0hy-1t8

CHAPTER 4: From Schrödinger's Equation to Quantum Conspiracy. Narrated by Ann G. Yu

https://youtu.be/0b-45KxvpJo

CHAPTER 5: The Limits of Einstein's Theory of Relativity. Narrated by Edward H. Yu

https://youtu.be/HVJkl9V7tGk

CHAPTER 6: Quantum Qubit Information Conspiracy. Narrated by Ann G. Yu

https://youtu.be/u6WKA1TiL94

CHAPTER 7: Why modern physics is so weird… and so wrong. Narrated by Edward H. Yu
   https://youtu.be/PiBxdhA03pM

CHAPTER 8: Einstein's General Theory Belong to the Realm of Science? Narrated by Edward H. Yu
   https://youtu.be/6S8YxnbBe-Q

CHAPTER 9: Einstein's Spooky Distance. Narrated by Warren Schlueter
   https://youtu.be/zpQL-E9vqc4

CHAPTER 10: Can Space Really Curves Spacetime? Narrated by Edward H. Yu
   https://youtu.be/Q3_IEdzAAyw

CHAPTER 11: Dark-Age of Our Modern Physics. Narrated by Warren Schlueter
   https://youtu.be/L2VqWJ7fOtc

CHAPTER 12: Where Dark Matter and Dark Energy comes from? Narrated by Warren Schlueter
   https://youtu.be/LNvqlaSYD14

CHAPTER 13: Why Einstein's relativistic mechanics is against the laws of nature. Narrated by Mary Urbach
   https://youtu.be/Sk7ZEg68V-o

CHAPTER 14: Myth of Entropy - Boltzmann's Exorcist. Narrated by Warren Schlueter
   https://youtu.be/DLBdSxP4N94

CHAPTER 15: Schematically Disproves Bogus Modern Principles. Narrated by Warren Schlueter
https://youtu.be/fmHWs8nLdlM

CHAPTER 16: From classical to physical realizable Hamiltonian Mechanics. Narrated by Warren Schlueter
https://youtu.be/G8RD5KdDKDE

CHAPTER 17: Enigma of Eigen state wave function. Narrated by Warren Schlueter
https://youtu.be/RlMzmzH2Ilk

CHAPTER 18: Why mathematics is not science. Narrated by Warren Schlueter
https://youtu.be/bVRHisMJQco

CHAPTER 19: Origin of our science. Narrated by Mary Urbach
https://youtu.be/Qe9ZPEvkmrg

CHAPTER 20: Origin of de Broglie wave dynamics. Narrated by Warren Schlueter
https://youtu.be/4U83X693Qa0

CHAPTER 21: Interpretation of $\psi(t)$. Narrated by Connie Etheridge
https://youtu.be/hhWYz55rGkE

CHAPTER 22: Why ancient Chinese did not have a fundamental science. Narrated by Warren Schlueter
https://youtu.be/h65QE-GWBSA

CHAPTER 23: AI at different level of knowledge of abstraction. Narrated by Tibor G. Varga

https://youtu.be/QTaL5xYLnSU

CHAPTER 24: Theory of temporal (t 0) Space. Narrated by Connie Etheridge.

https://youtu.be/kvGmmsbbyyM

CHAPTER 25: Dirac's antimatter hypothesis is not science. Narrated by Benjimen Yu

https://youtu.be/lNJsh5kZMVA

ABOUT THE AUTHOR

FRANCIS T. S. YU is an IEEE Fellow, SPIE Fellow, Evan Pugh Professor Emeritus of Electrical Engineering, Penn State University, and recipient of the Dennis Gabor Award ('04) and Emmett N. Leith Medal ('16).

He is the founding director of the Electro-Optics Lab at Penn State University and has authored 20 books (4 has coauthored), 4 monographs, 25 books chapters, over 300 refereed papers, over 250 conferences articles and 2 SPIE Milestone Series Works. He has co-edited over 25 SPIE Conference Proceedings. Some of his books have been translated into Russian, Chinese, Japanese and Korean.